It's another great book from CGP...

This book is packed with tricky practice questions to help you
make sure you've mastered Maths at KS3 (ages 11-14).

It's ideal if you're working at a higher level — it covers what would
have been called Levels 5-8 in the old Curriculum.

CGP — still the best! ☺

Our sole aim here at CGP is to produce the highest quality books —
carefully written, immaculately presented and dangerously close to being funny.

Then we work our socks off to get them out to you
— at the cheapest possible prices.

Contents

Section 1 — Numbers

Calculating Tips..1
Ordering Numbers..2
Addition and Subtraction...3
Multiplying Without a Calculator..................................4
Dividing Without a Calculator......................................5
Special Types of Number...6
Multiples and Factors...7
Primes and Prime Factors...8
Fractions, Decimals and Percentages..............................9
Fractions..10
Percentage Basics..11
Rounding Numbers..13
Rounding Errors and Estimating.................................14
Powers..15
Square Roots and Cube Roots....................................16
Standard Form..17

Section 2 — Algebra

Algebra — Simplifying Terms......................................18
Algebra — Expanding and Factorising.........................19
Solving Equations..20
Using Formulas..21
Expressions and Formulas from Words........................22
Equations and Formulas from Words...........................23
Rearranging Formulas..24
Number Patterns and Sequences.................................25
Inequalities...27

Section 3 — Graphs

X and Y Coordinates..28
Plotting Straight Line Graphs.....................................29
Gradients and $y = mx + c$...30
Real-Life Graphs...32
Reading Off Graphs...34
Quadratic Graphs..35

Section 4 — Ratio, Proportion & Rates of Change

Ratios 36
Direct Proportion 38
Inverse Proportion 39
Percentage Change 40
Converting Units 42
Maps and Scale Drawings 44
Best Buy 45
Density and Speed 46

Section 5 — Geometry and Measures

Symmetry 47
2D Shapes 48
Perimeter and Area 49
Circles 51
3D Shapes and Nets 52
Surface Area 53
Volume 54
Angle Basics 55
Geometry Rules 56
Interior and Exterior Angles 58
Transformations 60
Congruence and Similarity 62
Constructions 63
Pythagoras' Theorem 64
Trigonometry 65

Section 6 — Probability and Statistics

Probability 67
Venn Diagrams 69
Graphs and Charts 70
Mean, Median, Mode and Range 72
Averages from Frequency Tables 73
Scatter Graphs 76

Answers 78

Published by CGP

Editors:
Rob Harrison, Ruth Wilbourne

Contributors:
Deborah Dobson

With thanks to Simon Little and Mark Moody for the proofreading.

ISBN: 978 1 84146 038 3

Clipart from Corel®.
Printed by Elanders Ltd, Newcastle upon Tyne.

Based on the classic CGP style created by Richard Parsons

Calculating Tips

William Home Top tip

Here you go, your first page of lovely maths.   All of the stuff on this page comes down to good old **BODMAS**, so remember — **BODMAS, BODMAS, BODMAS, BODMAS, BODMAS...**

Q1   Edward uses his calculator to do this calculation: $\dfrac{12}{4 \times 0.5}$

a)   He gets an answer of 1.5. Is this correct?

b)   What has he done wrong?

c)   Write down the keys he should use to get the correct answer.

d)   Can you get the correct answer using another set of keystrokes?

1.5

Q2   Now Edward tries the following calculation, and gets an answer of 33.

$$\frac{140}{7 + 13}$$

What should he do to get the correct answer now?

Q3   Use your calculator to work these out:

(i) Writing down all the intermediate stages,
(ii) without writing down any intermediate stages.

a) $0.7 + (1.8 + 3.4) - (1.4 + 0.7)$

b) $8.2 - (4.1 + 1.6) - (0.7 - 3.7)$

c) $23.7 - 2 \times (4.3 - 1.9)$

d) $104 - 7 \times (3.2 - 11)$

e) $\dfrac{4.8 + 7.2}{0.2 \times 0.4}$

f) $\dfrac{37 - (21 - 4)}{4 \times \text{-}5}$

g) $3 \times (4 - 2 \times (0.7 \times 0.5))$

h) $\dfrac{2 \times 0.4^2 - 2 \times 0.2^2}{3.1 - 2.48}$

Remember:

> **Use BODMAS to get the order of operations right:**
>
> **Brackets,   Other,   Division,   Multiplication,   Addition,   Subtraction.**
>
> *First* ————————————————————————➤ *Last*

Q4   **Challenge** — by inserting as many brackets as you like,
see how many different answers you can get for the following:

$1 \times 3 + 5 - 3 \times 2 + 6 =$

Example: $((1 \times 3) + (5 - 3)) \times (2 + 6) = 40$

Ordering Numbers

Ideally, what you want to be able to do is order numbers without even thinking about it, so if you see a group of numbers you know the order of them straight away. Unfortunately it takes practice, but once it's done, it's done.

Q1 What are the largest and the smallest numbers that can be made with these sets of digits? Write each number out in words.

a) 4, 7, 9, 1

b) 3, 8, 8, 4

c) 3, 2, 4, 9

d) 5, 4, 3, 4, 8

e) 1, 2, 3, 7, 8

f) 1, 2, 3, 7, 9

Q2 What value does the digit 8 represent in each of these numbers?

a) 548.9

b) 784.2

c) 76.8

d) 4.081

e) 86560

f) 9.548

g) 7801

h) 823456

i) 18450

What a day...

Q3 Put these numbers in order, from the smallest to the largest.

a) 1.54   1.71   1.98   1.3   1.89   1.5   1.62

b) 102.8   101.2   100.3   102.89   100.4   101.6   100.43

c) 4   0   -1   -10   2   5   -3

d) 7.41   7.36   7.13   7.09   7.40   7.18   7.21

Q4 Put these measurements in descending order.

a) 4.0 cm   4.1 cm   2.3 cm   3.1 cm   2 cm   3.9 cm   0.9 cm

b) 76.1 km   79.1 km   74.9 km   74.1 km   75.2 km   78.7 km   74.3 km

c) 0.102 m   0.219 m   0.02 m   0.009 m   0.021 m   0.012 m   0.220 m

d) 40.73 g   40.93 g   40.81 g   41.06 g   40.07 g   41.1 g   40.7 g

Don't be put off by the units — as long as you're ordering numbers with the same unit just carry on as normal.

Addition and Subtraction

You're not allowed to use calculators on this page. And that's not 'cos I've got anything against them — it's just that far, far away in the future you'll have to do some non-calculator exams. Sorry about that...

NO CALCULATORS!!!!!

Q1 Use a pencil and paper to work out these calculations:

a) 1279 + 334

b) 4796 + 209

c) 569 – 491

d) 243 + 694 + 101

e) 3712 + 1319 + 2240

f) 7348 – 69

g) 1234 + 567 + 89

h) 9876 + 543 + 21

Q2 Calculate the following:

a) 2 – 7

b) -6 – 8

c) 0 – 9

d) 5 – -2

e) 7 – - 6

f) -8 – -2

g) -3 – -3

h) 8 – 5 – - 3

Q3 Using pencil and paper only, work out:

a) 31.8 + 42.7 + 83.8

b) 27.41 + 28.3 + 15.09

c) 2.31 + 23.1 + 231

d) 1046 + 164 + 0.146

e) 27 + 36 – 42 + 0.5

f) 234 – 34.2 + 4.23

g) 67.1 + 30.23 + 11.131 – 42.22

h) 0.012 + 0.314 + 0.505

There's no need to panic about sums involving decimals. Just make sure you get those decimal points lined up.

Q4 Work out these using paper and pencil only:

a) 47.0179 + 107.08 + 302.018

b) 73.179 + 8.987 + 20.117

c) 6.432 + 64.32 + 0.6432 + 643.2

d) -0.0002 + 0.0014 + 0.00024

e) 10.9 + -7.31

f) 173.7 + -87.89

Multiplying Without a Calculator

Some of these are pretty hard considering you can't use a calculator. But if you're struggling, there's a sure-fire way to get better at this kind of question — and that's to practise until you can do them.

NO CALCULATORS!!!!!

Q1 Carry out the following multiplications:

a) 51 × 10

b) 320 × 10

c) 14 × 100

d) 160 × 1000

e) 7.6 × 100

f) 5.487 × 10

Q2 Now work out these.

a) 43 × 47

b) 242 × 65

c) 721 × 341

d) 602 × 407

e) 34.7 × 2.3

f) 4.3 × 12.5

g) 73 × -0.14

h) 57.1 × -0.23

i) -4300 × -1.23

j) 3.12 × 8.33

Q3 Work out 87 × 231. Then use your answer to work out the following.

a) 8.7 × 2.31

b) 87 × 23.1

c) 0.87 × 231

d) 870 × 2.31

e) 8.7 × 23.1

f) 870 × 0.231

g) 0.087 × 2310

h) 8.7 × 0.231

Once you've got the first multiplication sorted, you can work out the rest by carefully shifting the decimal point.

Q4 Snailpace Coach Company is running a trip to "The Anoraks" concert. They have seven 52-seater coaches and eight 12-seater minibuses.

a) How many fans can they carry to the concert?

b) The coach company forks out £23 for each ticket. It also costs them £150 to run each coach and £80 per minibus. How much will the company have to pay in total to run the trip?

c) If they sell seats for the trip at £35 per ticket, how much profit will they make?

Almost there...

SNAILPACE COACHES

Dividing Without a Calculator

Divisions can be pretty easy when you're dividing by 10, 100 or 1000... but make sure that you know how to use short and long division for the harder ones.

NO CALCULATORS!!!!!

Q1   Carry out the following divisions:

a) 350 ÷ 10

b) 1500 ÷ 100

c) 190,000 ÷ 1000

d) 20 ÷ 100

e) 1.6 ÷ 10

f) 410.36 ÷ 100

Q2   Work out the following, without using a calculator.

a) 357 ÷ 7

b) 744 ÷ 3

c) 676 ÷ 4

d) 276 ÷ 23

e) 231 ÷ 21

f) 437 ÷ 19

g) 8.7 ÷ 0.3

h) 48.96 ÷ 0.06

> To get rid of decimals, write the division as a fraction and multiply both the top and bottom by 10 or 100.

Q3   Give these answers as a whole number plus remainder.

a) 985 ÷ 4

b) 767 ÷ 3

c) 371 ÷ 6

d) 423 ÷ 13

e) 279 ÷ 23

f) 986 ÷ 46

g) 779 ÷ 37

h) 775 ÷ 15

Q4   Cedric breeds rats. He keeps 7 rats in each cage.

How many cages will he need for 81 rats?

Q5   Daisy breeds locusts. How many 5 kg bags of food will she need to last four weeks if her locusts eat 1 kg of food a day?

Q6   Cedric suggests feeding the locusts to his rats.

If each rat eats 4 locusts a day, how long will 3416 locusts last 7 rats?

Special Types of Number

These can sound pretty complicated, but it's mainly just a matter of knowing what the words mean. Unfortunately, that means you're gonna have to learn them. Doh.

Q1   Match these four number sequences with their names:

a)   2, 4, 6, 8, …

b)   1, 3, 5, 7, …

c)   1, 4, 9, 16, …

d)   1, 8, 27, 64, …

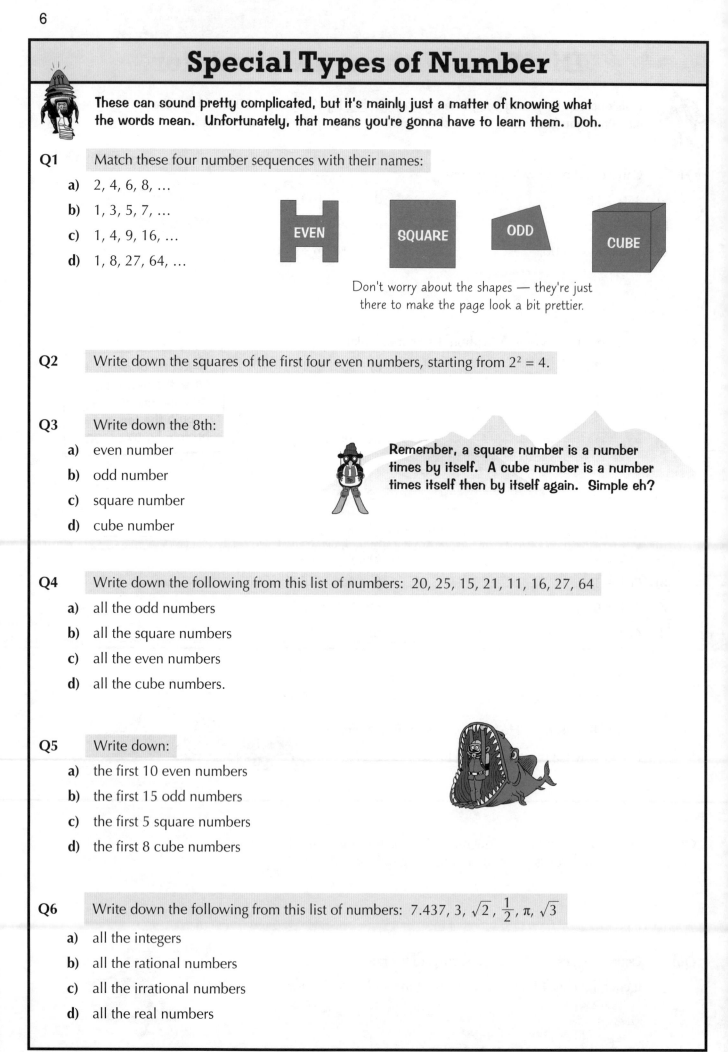

Don't worry about the shapes — they're just there to make the page look a bit prettier.

Q2   Write down the squares of the first four even numbers, starting from $2^2 = 4$.

Q3   Write down the 8th:

a)   even number

b)   odd number

c)   square number

d)   cube number

Remember, a square number is a number times by itself. A cube number is a number times itself then by itself again. Simple eh?

Q4   Write down the following from this list of numbers: 20, 25, 15, 21, 11, 16, 27, 64

a)   all the odd numbers

b)   all the square numbers

c)   all the even numbers

d)   all the cube numbers.

Q5   Write down:

a)   the first 10 even numbers

b)   the first 15 odd numbers

c)   the first 5 square numbers

d)   the first 8 cube numbers

Q6   Write down the following from this list of numbers: $7.437, 3, \sqrt{2}, \frac{1}{2}, \pi, \sqrt{3}$

a)   all the integers

b)   all the rational numbers

c)   all the irrational numbers

d)   all the real numbers

Multiples and Factors

Q1   Sort these numbers into 3 lists: multiples of 3, multiples of 4 and multiples of 5.
33   25   1016   164   21   63   10   39   175   50   4036   51   35   11144   110   512

    a)   In the multiples of 5, what do you notice about the last digit?

    b)   In the multiples of 3, what do you notice about the digit sum?

    c)   In the multiples of 4, what do you notice about the last 2 digits?

> For the last bit, you need to look at the 2 digit number at the end — what does it always divide by?

Q2   Is 3 a factor of 2001?

Q3   What's the LCM of 8 and 12?

Q4   Write down:

    a)   the first 12 multiples of 6, and the first 10 multiples of 8.

    b)   any common multiples (the ones that are in both lists).

    c)   the lowest common multiple (LCM).

Q5   Find:

    a)   all the factors, in order, of each of these numbers: 12   18   24   30

    b)   the common factors.

    c)   the highest common factor (HCF) of the four numbers.

Q6   Find the highest common factor of the following sets of numbers:

    a)   32 and 48.

    b)   45 and 105.

    c)   36, 84 and 132.

Q7   Find all the factors of 300.

Q8   Craggy Point Lighthouse flashes every 25 seconds, and Devil's Rock Lighthouse flashes every 40 seconds. If they flash together, how soon will it be before they flash together again?

Who the devil's that?

These 2 are asking more or less the same question — you need to find the lowest common multiple of both timespans.

Q9   There are two sets of traffic lights outside Eric's house. One day, he times how often they change. Set A turns green every 60 seconds. Set B turns green every 70 seconds. At midday precisely they both turn green together. At what time will they both turn green together again?

Primes and Prime Factors

Q1 In the ten by ten square opposite, circle all the prime numbers. The first three have been done for you.

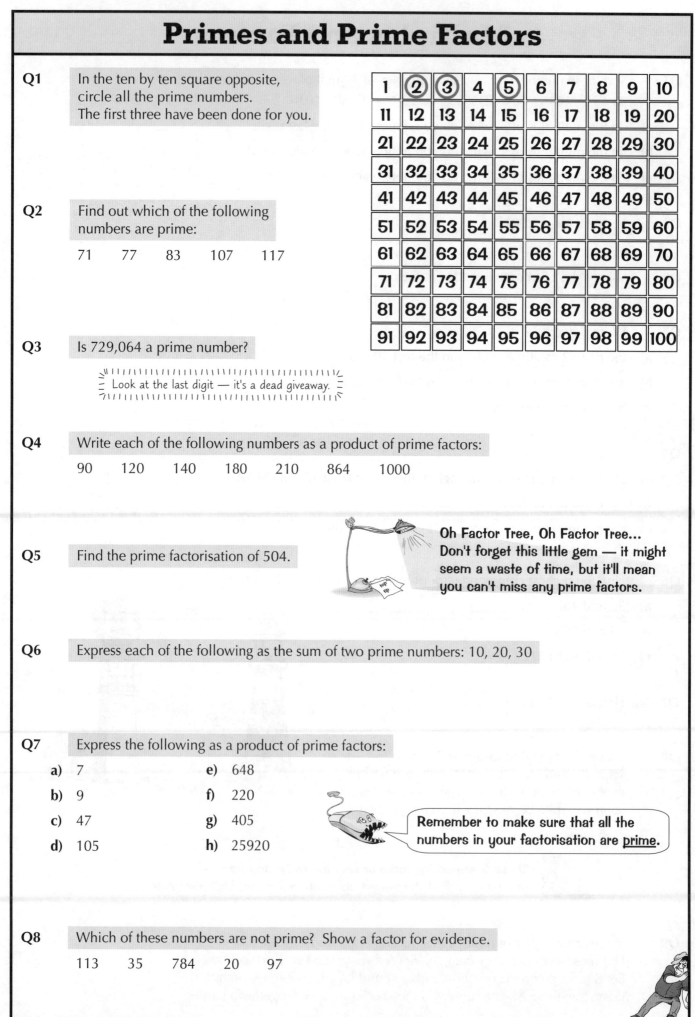

1	②	③	4	⑤	6	7	8	9	10
11	12	13	14	15	16	17	18	19	20
21	22	23	24	25	26	27	28	29	30
31	32	33	34	35	36	37	38	39	40
41	42	43	44	45	46	47	48	49	50
51	52	53	54	55	56	57	58	59	60
61	62	63	64	65	66	67	68	69	70
71	72	73	74	75	76	77	78	79	80
81	82	83	84	85	86	87	88	89	90
91	92	93	94	95	96	97	98	99	100

Q2 Find out which of the following numbers are prime:

71   77   83   107   117

Q3 Is 729,064 a prime number?

Look at the last digit — it's a dead giveaway.

Q4 Write each of the following numbers as a product of prime factors:

90   120   140   180   210   864   1000

Q5 Find the prime factorisation of 504.

Oh Factor Tree, Oh Factor Tree... Don't forget this little gem — it might seem a waste of time, but it'll mean you can't miss any prime factors.

Q6 Express each of the following as the sum of two prime numbers: 10, 20, 30

Q7 Express the following as a product of prime factors:

a) 7   e) 648

b) 9   f) 220

c) 47   g) 405

d) 105   h) 25920

Remember to make sure that all the numbers in your factorisation are <u>prime</u>.

Q8 Which of these numbers are not prime? Show a factor for evidence.

113   35   784   20   97

Fractions, Decimals and Percentages

The main thing to remember is that there's no real difference between fractions, decimals and percentages — they all just mean "a part of..."

Q1 Change these decimals to percentages:

a) 0.28

b) 0.57

c) 0.875

d) 0.4725

e) 0.04

f) 0.045

These are the easy ones — all you've got to do is move the decimal point.

Q2 Change these percentages to decimals:

a) 35%

b) 35.8%

c) 70%

d) 7%

e) 0.7%

f) 5.5%

Q3 Turn these fractions into exact percentages:

a) $\frac{7}{8}$

b) $\frac{5}{16}$

c) $\frac{13}{40}$

d) $\frac{17}{20}$

e) $\frac{14}{25}$

f) $\frac{67}{80}$

Q4 Work these out as decimals correct to 3 decimal places, then write them as percentages:

a) $\frac{2}{9}$

b) $\frac{13}{15}$

c) $\frac{7}{18}$

d) $\frac{4}{11}$

e) $\frac{7}{12}$

f) $\frac{5}{13}$

Q5 Turn these percentages into fractions in their lowest terms:

a) 12.5%

b) 37.5%

c) 62.5%

d) 87.5%

e) 7.5%

f) 17.5%

Always do fractions to decimals to percentages and back in that order. It's a good reliable method to stick to.

Q6 Tariq got these scores in his exams:

$\frac{37}{50}$ for English, $\frac{14}{20}$ for History, $\frac{71}{85}$ for Maths and $\frac{39}{60}$ for Basket Weaving.

a) Convert these marks to percentages (to 1 decimal place if necessary).

b) Which were his best and worst results?

Q7 Jamila has scored 18 out of 25 in her test, and Diana has scored 16 out of 20. Who got the higher percentage?

Fractions

Don't forget to make the bottom numbers the same when adding or subtracting fractions. Avoid using your calculator for these questions (unless you're told otherwise).

Q1 For each fraction pair, put them both over a common denominator to see which is bigger. Write out the original fraction pair using the "greater than" sign $>$:

a) $\frac{3}{4}, \frac{4}{5}$

b) $\frac{2}{3}, \frac{5}{8}$

c) $\frac{1}{3}, \frac{2}{5}$

d) $\frac{13}{20}, \frac{7}{10}$

Q2 Add these fractions together, giving your answers in their simplest forms:

a) $\frac{1}{5} + \frac{1}{5}$

b) $\frac{1}{4} + \frac{3}{4}$

c) $\frac{3}{16} + \frac{5}{16}$

d) $1\frac{1}{4} + \frac{3}{4}$

e) $1\frac{2}{5} + \frac{4}{5}$

f) $2\frac{2}{3} + \frac{2}{3}$

g) $\frac{7}{2} + \frac{7}{3}$

h) $\frac{14}{3} + \frac{23}{4}$

Q3 Do these subtractions, giving your answers in their simplest forms:

a) $\frac{7}{8} - \frac{1}{8}$

b) $\frac{4}{5} - \frac{2}{5}$

c) $\frac{2}{3} - \frac{1}{3}$

d) $1 - \frac{2}{5}$

e) $2 - \frac{5}{8}$

f) $1\frac{3}{4} - \frac{1}{4}$

g) $\frac{23}{8} - \frac{5}{8}$

h) $\frac{9}{5} - \frac{3}{5}$

Q4 Find:

a) $\frac{1}{5}$ of £80

b) $\frac{1}{8}$ of 48 kg

c) $\frac{1}{6}$ of 24 000 people

d) $\frac{1}{4}$ of 180 days

e) $\frac{2}{3}$ of 90°

f) $\frac{3}{5}$ of £4500

Q5 Do these first by hand, then check the results by calculator:

a) $\frac{3}{4} \times \frac{3}{4}$

b) $\frac{2}{5} \times \frac{1}{8}$

c) $1\frac{7}{8} \times 1\frac{4}{5}$

d) $3\frac{5}{8} \times 2\frac{2}{15}$

Q6 Do these first by hand, then by calculator and compare:

a) $\frac{2}{5} \div \frac{1}{4}$

b) $1\frac{3}{4} \div \frac{5}{8}$

c) $3\frac{1}{2} \div 2\frac{1}{4}$

d) $5\frac{1}{3} \div 2\frac{2}{3}$

Q7 Farmers in Broughton have been growing vegetables.

a) Farmer Richard owns 900 hectares of land. At the moment $\frac{2}{5}$ of this is used for growing vegetables. How many hectares is this?

b) Farmer Paddy down the road has a 1200 hectare farm, of which 720 hectares are for vegetables. What fraction of his land is for vegetables? Give your answer in its simplest form.

c) Farmer Richard wants to increase his production of vegetables so that it uses the same fraction of his land as Farmer Paddy. How many more hectares must he turn over to vegetables?

Percentage Basics

There's a treat in store for you now — two whole pages of lovely percentage questions. This first one's all about finding a percentage of a number, so when you see a % symbol you know what to do — write the percentage as a decimal and then multiply.

Q1 How much is:

a) 1% of £35
b) 2% of £18
c) 5% of £90

d) 8% of £60
e) 22% of £4
f) 35% of £2

g) 60% of £5
h) 150% of £200
i) 125% of £500?

Q2 What is:

a) 16% of £3200
b) 20% V.A.T. on a bill of £48.80
c) 27% of 550 square miles

d) 92% of 6500 people
e) 16% of 350 lizards
f) 18% of 2250 cars?

Q3 Rewrite, translating the percentages into actual numbers:

a) Out of 44500 voters in the town, 32% voted for the Conservatives.
b) 18% of 3500 cars stopped had defects.
c) 15% of the cake's weight of 450 grams is butter.
d) 19% of the 1300 rare birds found were diseased.

Q4 Find:

a) 4% of 550 children
b) 7% of 900 grams
c) 2% of 4500 lorries
d) 8% of 2550 insects
e) 3% of 1400

f) 5% of 6500
g) 7.5% of £30
h) 2.5% of £4200
i) 100.5% of 360
j) 120.5% of 2000 grams

Q5 Is 34% of 68 the same as 68% of 34?
Is 5% of 43 the same as 43% of 5?
Will this always be true?

Percentage Basics

...and now for your second percentage instalment. This page gives you tons of practice at finding a number as a percentage of another number. It's not as bad as it looks — to find x as a percentage of y, divide x by y then multiply by 100.

Q6   Rewrite the following sentences, using percentages:

a)   3 out of every 10 people in Darkley believe in ghosts.

b)   4 out of 5 people are against annoying ringtones on buses.

c)   One in every eight workers are off sick at present.

d)   Only 3 out of 20 children thought there should be more homework.

Q7   580 books were delivered to a warehouse and 29 were found to be damaged. What percentage were damaged?

Q8   A 150 gram serving of fruit salad contains 17 grams of sugar. What percentage is sugar? (1 decimal place)

Q9   250 g of butter contains 202 g of fat. What percentage is fat?

Q10   Rocky Canyon Mine can't produce copper economically unless the ore contains at least 28% of the metal. Recently 4500 tonnes of ore has yielded 1168 tonnes of copper. Can they carry on?

These questions can have a lot of waffle in 'em... and most of it you can just ignore — put the story into maths, then forget about the rest of it.

Q11   Sarah has bought herself a new laptop costing £1250. She's also bought a new printer for £150 and a desk for £100.

a)   How much did she spend altogether?

b)   What percentage of her total outlay was the cost of the printer?

c)   What percentage of her total outlay was the cost of the laptop?

Rounding Numbers

Well, this page should give you plenty of practice at rounding — and isn't that just what you've always wanted...?

Q1 Round to one decimal place:

a) 4.73   c) 6.75   e) 11.76

b) 8.92   d) 19.476   f) 20.85

Q2 Round to two decimal places:

a) 4.763   c) 17.094   e) 14.986

b) 5.0852   d) 12.990   f) 17.098

Q3 Round these weights off to the nearest gram.

a) 4.86932 kg   c) 1.00982 kg   e) 3.0605 kg

b) 1.00942 kg   d) 2.070695 kg   f) 0.0039 kg

> Remember those units:
> 1 g = 0.001 kg

Q4 Round these angles to the nearest $\frac{1}{10}$ degree.

a) 12.83°   c) 27.04°   e) 57.8159°

b) 12.89°   d) 24.97°   f) 57.8951°

Q5 Round off these distances to the nearest 100 metres – i.e. to one decimal place:

a) 5.768 km   c) 8.48 km   e) 17.685 km

b) 9.039 km   d) 8.41 km   f) 17.658 km

Q6 Write these numbers correct to 3 significant figures:

a) 6762   c) 6769.5   e) 2009.75

b) 6767.5   d) 2005   f) 2000

Q7 Write these numbers correct to 2 significant figures:

a) 0.352   c) 0.00574   e) 0.0356

b) 0.0357   d) 4.01964   f) 1.0356

Q8 Write the populations of these cities to 3 significant figures, which in this particular case will be the same as rounding off to the nearest 1000:

a) Bigtown – 369387   d) Littlewich – 129960

b) Shortville – 102008   e) Megaborough – 479940

c) Middlethorpe – 190886   f) Port Average – 157095

Rounding Errors and Estimating

Q1 What is the error when each of these numbers is given to 1 significant figure?

a) 7.2

b) 8.4

c) 10.21

d) 7.56

e) 8888

f) 13012

Q2 Find the range of possible values for x for each of the following. Give your answers as inequalities.

a) $x = 120$ to the nearest 10

b) $x = 300$ to the nearest 100

c) $x = 10.2$ to 1 d.p.

d) $x = 7.5$ to 2 s.f.

e) $x = 8800$ to 2 s.f.

f) $x = 1010$ to 3 s.f.

Q3 For each calculation: **i)** round off the figures to 1 significant figure, and work out an estimate to the calculation.
ii) use your calculator to find a more exact answer. Round it to 3 s.f.

a) $6.81 + 9.13 + 17.93$

b) $63.56 - 42.85$

c) 8.63×7.42

d) $\dfrac{4.35 \times 2.86}{1.92}$

e) $\dfrac{91.2 - 72.4}{17.68}$

f) $\dfrac{99.8 \times 4.7}{9.84}$

Remember — round everything to 1 sig fig... then do the calculation.

Q4 **i)** Round off the figures to 1 significant figure, and use them to estimate the answer to the calculation.
ii) Use your calculator to find a more exact answer, to 3 d.p.:

a) $\dfrac{0.38 \times 1.14}{0.189}$

b) $\dfrac{3.725 - 1.628}{4.96 \times 1.98}$

c) $\dfrac{1.12 \times 0.880}{1.08 \times 2.970}$

d) $0.59 + 1.42 - 0.385$

e) $\dfrac{0.803}{3.965} + 1.074$

f) $\dfrac{5.843 + 8.925 - 3.185}{7.24 - 2.19}$

Always show your working — you sometimes get marks for that even if your answer's wrong.

Q5 A car goes 407 km in 5.11 hours.

a) By rounding off to 1 significant figure, give a rough estimate of its speed in kilometres per hour.

b) Use your calculator to find a more exact result to 3 significant figures.

Powers

Unfortunately, maths powers aren't quite as fun as the superhero kind... but they do obey nifty rules. When you're multiplying, add the powers, and when you're dividing, subtract them.

Q1 Work out the exact value of:

a) 2^5 d) 2^8 g) 10^5 j) 6^3

b) 3^3 e) 3^4 h) 100^3 k) 7^3

c) 4^2 f) 5^3 i) 8^3 l) 10^6

Q2 Simplify by adding or subtracting powers; then work out the exact value:

a) $4^2 \times 4^3$ d) $2^7 \times 2^4$ g) $5^6 \div 5^4$

b) $2^3 \times 2^5$ e) $3^7 \div 3^5$ h) $7^{10} \div 7^9$

c) $3^6 \times 3^3$ f) $10^{12} \div 10^9$ i) $4^6 \div 4^3$

Q3 Simplify as far as possible (in some cases, this just means removing the × signs):

a) $a \times a \times a$ e) $x \times y$ i) $a \times b \times c \times 5$

b) $2 \times a \times a \times a$ f) $x \times y \times z$ j) $3 \times x \times x \times 4 \times y$

c) $3 \times 2 \times x \times x \times x$ g) $x \times x \times x \times y$ k) $2 \times y \times x \times 4$

d) $5 \times y \times 4 \times y$ h) $x \times x \times y \times y \times y$ l) $10 \times k \times j \times k \times j$

Q4 Simplify using the power rules:

a) $x^{10} \div x^4$ c) $a^7 \div a^4$ e) $\dfrac{r^5}{r}$

b) $y^5 \div y^2$ d) $\dfrac{b^6}{b^3}$ f) $\dfrac{y^{10}}{y^7}$

Q5 Use the power rules to simplify the following:

a) $3a \times 5a \times 4$ d) $2a^3 \times 3a^2$ g) $(x^2)^2$

b) $12x \times 3x^2$ e) $3p \times 2p^2 \times 4p^3$ h) $(y^4)^3$

c) $4y^2 \times 5y$ f) $7m^2 \times 3n$ i) $(x^3)^{-2}$

Q6 Simplify using the power rules:

a) $\dfrac{10x^4}{5x^3}$ b) $\dfrac{15a^5}{3a^2}$ c) $\dfrac{12b^4}{4b^3}$ d) $\dfrac{20k^5}{5k^2}$ e) $\dfrac{27x^5}{18y^5}$

Q7 Write using negative powers:

a) $\dfrac{1}{10^2}$ b) $\dfrac{1}{x^2}$ c) $\dfrac{1}{10^4}$ d) $\dfrac{1}{a^4}$ e) $\dfrac{5}{a^4}$

Square Roots and Cube Roots

Those weird tick signs and microscopic numbers are nowhere near as scary once you know what they mean. $\sqrt{20}$ is the "square root of 20", which is the <u>number which times by itself gives 20</u>. $\sqrt[3]{20}$ means the "cube root of 20" — that's the number for which <u>number × number × number = 20</u>.

Q1   Use your calculator to find (to 2 dp):

   a) $\sqrt{50}$ **c)** $\sqrt{65}$ **e)** $\sqrt{7}$

   b) $\sqrt{20}$ **d)** $\sqrt{15}$ **f)** $\sqrt{72}$

Q2   Find the following to 1 dp:

   a) $\sqrt[3]{80}$ **d)** $\sqrt[4]{75}$

   b) $\sqrt[3]{150}$ **e)** $\sqrt[3]{63}$

   c) $\sqrt[4]{5}$ **f)** $\sqrt[3]{10}$

Q3   Find both square roots of the following numbers:

   a)   49   **b)**   256   **c)**   90.25   **d)**   86.49

Q4   Given y^2, write down the two possible values of y:

   a) $y^2 = 81$ **d)** $y^2 = 100$

   b) $y^2 = 25$ **e)** $y^2 = 4$

   c) $y^2 = 16$ **f)** $y^2 = 36$

Q5   Given y^3 write down the value of y:

   a) $y^3 = 125$ **d)** $y^3 = 27$

   b) $y^3 = 64$ **e)** $y^3 = 1$

   c) $y^3 = 8$ **f)** $y^3 = 0$

Check your answers using a <u>calculator</u> afterwards.

Q6   Simplify:

   a) $\sqrt{16x^2}$ **f)** $\sqrt{a^4}$

   b) $\sqrt{25a^2}$ **g)** $\sqrt[3]{27a^3}$

   c) $\sqrt{100m^2}$ **h)** $\sqrt[3]{64a^3b^3}$

   d) $\sqrt{64a^2b^2}$ **i)** $\sqrt[3]{1000a^6}$

   e) $\sqrt{16a^2b^2c^2}$ **j)** $\sqrt[3]{a^6}$

Standard Form

Writing very big (or very small) numbers gets a bit messy with all those zeros, if you don't use this standard index form. But of course, the main reason for knowing about standard form is... you guessed it — it could come up in a test.

1 billion = 1000 million

Q1 Write in standard form:

a) 5000
b) 9000
c) 90 000
d) 200 000
e) 3 million
f) 30 million
g) 300 million
h) 8 billion
i) 10 billion

Q2 Write in standard form:

a) 5 million
b) 5.8 million
c) 5.85 million
d) 6 000 000
e) 6 700 000
f) 6 750 000

Q3 Write as ordinary numbers:

a) 4×10^3
b) 4.3×10^3
c) 4.35×10^3
d) 4.352×10^5
e) 6×10^4
f) 6.4×10^4
g) 6.42×10^4
h) 6.425×10^4
i) 6.4258×10^4

Q4 Write these numbers in correct standard form, i.e. with just one digit before the point:

a) 35×10^5
b) 160×10^3
c) 45×10^6
d) 127×10^6
e) 58.5×10^4
f) 72.8×10^9
g) 0.3×10^5
h) 0.85×10^6
i) 0.03×10^5

Q5 Evaluate in standard form:

a) $(3 \times 10^4) \times (2 \times 10^5)$
b) $(1.5 \times 10^6) \times (2 \times 10^4)$
c) $(4 \times 10^8) \times (3 \times 10^5)$
d) $(2.5 \times 10^5) \times (5 \times 10^4)$

Don't forget to use your power rules: When you multiply, add the powers. When you divide, subtract them.

Q6 Carry out the following divisions, giving your answers in standard form:

a) $(9 \times 10^7) \div (3 \times 10^2)$
b) $(8 \times 10^{12}) \div (2 \times 10^4)$
c) $(7 \times 10^9) \div (2 \times 10^3)$
d) $(6 \times 10^5) \div (3 \times 10^3)$

Q7 Write these numbers in standard form:

a) 0.0004
b) 0.02
c) 0.025
d) 0.0005
e) 0.00052
f) 0.000527

Q8 Put these in order from smallest to largest: 2.31×10^3, 2450, 1.76×10^3.

Q9 Put these in order from smallest to largest: 1.6×10^{-4}, 6.5×10^{-5}, 0.0078

Section 2 — Algebra

Algebra — Simplifying Terms

Algebra can be pretty scary at first. But don't panic — the secret is just
to practise lots and lots of questions. Eventually you'll be able to do it
without thinking, just like riding a bike. But a lot more fun, obviously...

Q1   Simplify these expressions.

a) $3z + 2y - z$

b) $5a - 3a + 6a$

c) $4w + 2v - 4v + 2w$

d) $5c - 3d - 2d$

e) $3 + t + 2u - 3t + 7$

f) $f - 7 + 3f - 2g + 8$

g) $3s + 2r - 7 + 3r - 7s + 4$

h) $3 - 2h - 3j + h - 4$

Q2   Simplify:

a) $a \times a$

b) $b \times b \times b$

c) $c \times c \times c \times c$

d) $a \times b \times a$

e) $2d \times f \times f \times d$

f) $g^2 \times g^2$

g) $2hg \times hg$

h) $3k^3 \times 4k$

The trick here is to collect
together letters that are
the same — use the
power rules to help you.

Q3   These are more tricky. Simplify:

a) $z^2 \div z$

b) $y^4 \div y$

c) $\dfrac{x^7}{x^3}$

d) $\dfrac{4k^2}{2k}$

e) $20m^4 \div 4m^3$

f) $0.5n^7 \div 0.25n^2$

g) $p \div p$

h) $\dfrac{q^3}{q^3}$

Q4   Match each expression on the left with an equal expression on the right.

$p + p$	$2p$
$a - b + a$	0
$r + s - r$	$2v - 3$
$(t + t) \div 4u$	s
$v + 2 + v - 5$	$\dfrac{t}{2u}$
$a \times b$	$2a - b$
$\dfrac{1}{2} \times w \times x$	$\dfrac{wx}{2}$
$3 + y - 7 - y + 4$	ba

Algebra — Expanding and Factorising

Another lovely page of algebra...
It might look a bit weird but believe me — it's bound to come up in a test.

Q1 Expand these expressions.

a) $2(p + 3)$

b) $3(q - 3)$

c) $4(4 + r)$

d) $5(2 - r)$

e) $3(2s + 1)$

f) $4(3s - 7)$

g) $5(10 + 3t)$

h) $6(-2 - 2u)$

Q2 Multiply out these brackets:

a) $3(a + b)$

b) $3(m + 2n + 5k)$

c) $6(2x - 3y)$

d) $10(5x - 4y + 6z)$

e) $-2(x + y)$

f) $-2(2c - 5d)$

g) $-8(-3a - 4b + 6)$

h) $2a(b + 3c)$

i) $-3m(x + y)$

j) $-3m(m + n)$

k) $4h(l - h)$

l) $6y(y^2m + n)$

Q3 Factorise these expressions:

a) $2a + 4$

b) $6b + 9$

c) $3c - 6$

d) $8 - 4d$

e) $2x + 2y$

f) $3x + 6y$

g) $ax + ay$

h) $10a + 15b$

i) $4x - 2y$

j) $6x - 9xz$

Remember that when you're factorising, pull out all the things that are in all terms, then put brackets round the rest...

Q4 Factorise these expressions:

a) $7p^2 - 14pq$

b) $a^2p - aq$

c) $3xy^2 + 3y^2 + 3yz$

d) $4ax + 8ay$

e) $12bx - 6by + 24bz$

f) $a - 4a^2$

g) $n^2 + 5n$

h) $x^2 - x$

Q5 Multiply out these brackets:

a) $(x + 4)(x + 2)$

b) $(x + 6)^2$

c) $(x - 3)(x + 5)$

d) $(x - 3)^2$

e) $(a + 3)(2a + 3)$

f) $(a - 3)(a + 4)$

g) $(m - 2)(m - 3)$

h) $(2m + 5)(m + 1)$

i) $(3y + 2)(y - 5)$

j) $(4x + 3)^2$

k) $(2k + 1)(k + 1)(k + 4)$

l) $(3y - 4)(y + 2)(y + 1)$

You'll need **FOIL** for these:

First   **O**utside   **I**nside   **L**ast

These last two are tricky. Start by multiplying out the first two brackets.

Solving Equations

 You don't need to be a super-sleuth to solve equations, but you will need practice. Always do the same thing to both sides of the equation, and you can't go far wrong. Just keep going 'til you've got the letter on its own.

Q1 Solve the following:

a) $4x = 20$

b) $7x = 28$

c) $x + 3 = 11$

d) $x + 19 = 23$

e) $x - 6 = 13$

f) $7x = -14$

g) $2x = -18$

h) $x + 5 = -3$

i) $\frac{x}{2} = 22$

j) $\frac{x}{7} = 3$

k) $\frac{x}{5} = 8$

l) $10x = 100$

m) $2x + 1 = 7$

n) $2x + 4 = 5$

 Check your answer by sticking it back into the equation at the end and seeing if it works.

Q2 Solve the following equations:

a) $3(2x + 5) = 39$

b) $7(x - 2) = 126$

c) $9(3x + 4) = 306$

d) $8(5x - 3) = 136$

e) $6(4x + 7) = 282$

f) $7(9x - 8) = 6244$

Q3 Solve:

a) $5x - 9 = 41$

b) $\frac{x}{7} + 14 = 20$

c) $\frac{3x}{4} - 9 = 6$

d) $11x + 4 = 6x + 29$

e) $3x + 8 + 4x - x = 26$

f) $\frac{2x}{3} = 10$

g) $2(3x - 5) = 170$

h) $\frac{4x}{5} - 8 = 72$

i) $10x - 9 - 3x = 40$

j) $x + 2x + 3x + 4x = 1000$

Q4 Solve the following:

a) $5(x - 1) + 3(x - 4) = -11$

b) $3(x + 2) + 2(x - 4) = x - 3(x + 3)$

c) $\frac{3x}{2} + 3 = x$

d) $3(4x + 2) = 2(2x - 1)$

e) $5x + \frac{7}{9} = 3$

f) $2x + \frac{7}{11} = 3$

Using Formulas

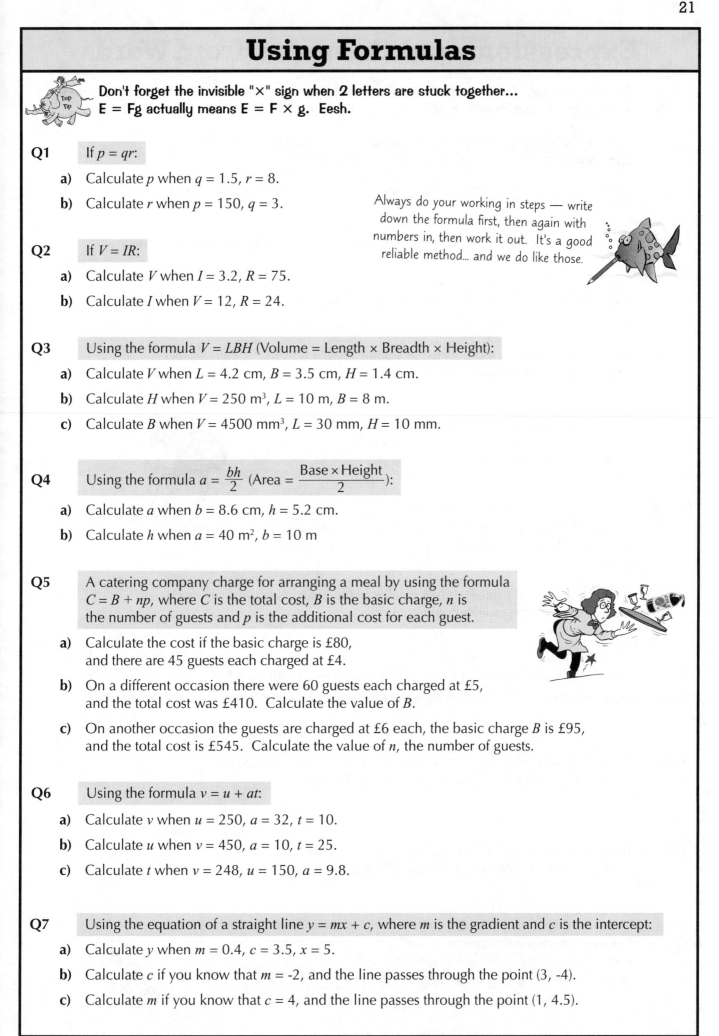

Don't forget the invisible "×" sign when 2 letters are stuck together...
E = Fg actually means E = F × g. Eesh.

Q1 If $p = qr$:

   a) Calculate p when $q = 1.5$, $r = 8$.

   b) Calculate r when $p = 150$, $q = 3$.

Always do your working in steps — write down the formula first, then again with numbers in, then work it out. It's a good reliable method... and we do like those.

Q2 If $V = IR$:

   a) Calculate V when $I = 3.2$, $R = 75$.

   b) Calculate I when $V = 12$, $R = 24$.

Q3 Using the formula $V = LBH$ (Volume = Length × Breadth × Height):

   a) Calculate V when $L = 4.2$ cm, $B = 3.5$ cm, $H = 1.4$ cm.

   b) Calculate H when $V = 250$ m³, $L = 10$ m, $B = 8$ m.

   c) Calculate B when $V = 4500$ mm³, $L = 30$ mm, $H = 10$ mm.

Q4 Using the formula $a = \dfrac{bh}{2}$ (Area = $\dfrac{\text{Base} \times \text{Height}}{2}$):

   a) Calculate a when $b = 8.6$ cm, $h = 5.2$ cm.

   b) Calculate h when $a = 40$ m², $b = 10$ m

Q5 A catering company charge for arranging a meal by using the formula $C = B + np$, where C is the total cost, B is the basic charge, n is the number of guests and p is the additional cost for each guest.

   a) Calculate the cost if the basic charge is £80, and there are 45 guests each charged at £4.

   b) On a different occasion there were 60 guests each charged at £5, and the total cost was £410. Calculate the value of B.

   c) On another occasion the guests are charged at £6 each, the basic charge B is £95, and the total cost is £545. Calculate the value of n, the number of guests.

Q6 Using the formula $v = u + at$:

   a) Calculate v when $u = 250$, $a = 32$, $t = 10$.

   b) Calculate u when $v = 450$, $a = 10$, $t = 25$.

   c) Calculate t when $v = 248$, $u = 150$, $a = 9.8$.

Q7 Using the equation of a straight line $y = mx + c$, where m is the gradient and c is the intercept:

   a) Calculate y when $m = 0.4$, $c = 3.5$, $x = 5$.

   b) Calculate c if you know that $m = -2$, and the line passes through the point (3, -4).

   c) Calculate m if you know that $c = 4$, and the line passes through the point (1, 4.5).

Expressions and Formulas from Words

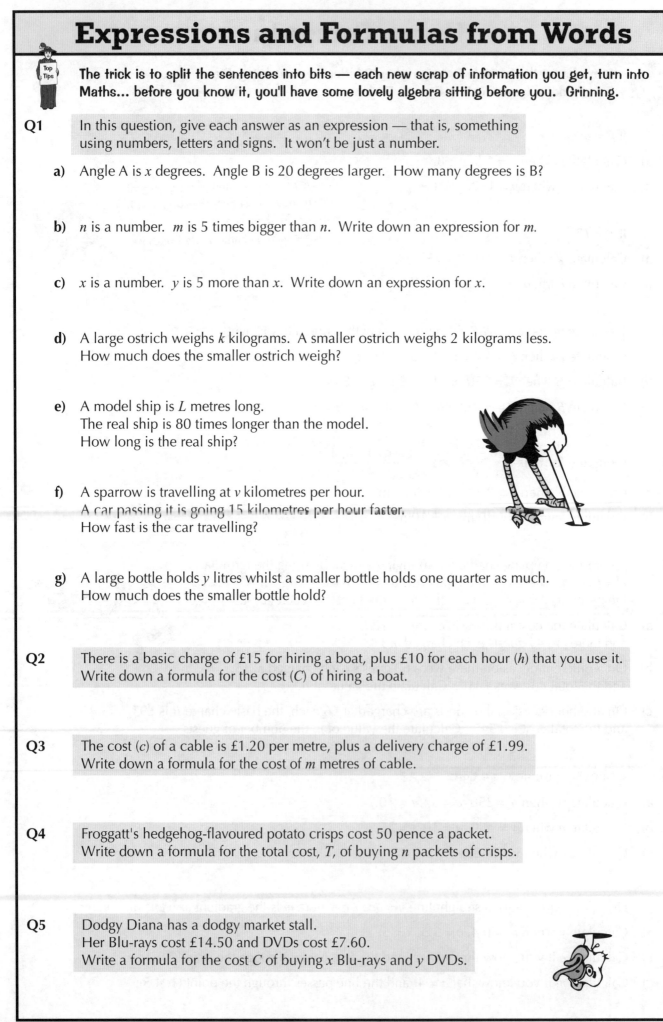

The trick is to split the sentences into bits — each new scrap of information you get, turn into Maths... before you know it, you'll have some lovely algebra sitting before you. Grinning.

Q1 In this question, give each answer as an expression — that is, something using numbers, letters and signs. It won't be just a number.

a) Angle A is x degrees. Angle B is 20 degrees larger. How many degrees is B?

b) n is a number. m is 5 times bigger than n. Write down an expression for m.

c) x is a number. y is 5 more than x. Write down an expression for x.

d) A large ostrich weighs k kilograms. A smaller ostrich weighs 2 kilograms less. How much does the smaller ostrich weigh?

e) A model ship is L metres long.
The real ship is 80 times longer than the model.
How long is the real ship?

f) A sparrow is travelling at v kilometres per hour.
A car passing it is going 15 kilometres per hour faster.
How fast is the car travelling?

g) A large bottle holds y litres whilst a smaller bottle holds one quarter as much.
How much does the smaller bottle hold?

Q2 There is a basic charge of £15 for hiring a boat, plus £10 for each hour (h) that you use it.
Write down a formula for the cost (C) of hiring a boat.

Q3 The cost (c) of a cable is £1.20 per metre, plus a delivery charge of £1.99.
Write down a formula for the cost of m metres of cable.

Q4 Froggatt's hedgehog-flavoured potato crisps cost 50 pence a packet.
Write down a formula for the total cost, T, of buying n packets of crisps.

Q5 Dodgy Diana has a dodgy market stall.
Her Blu-rays cost £14.50 and DVDs cost £7.60.
Write a formula for the cost C of buying x Blu-rays and y DVDs.

Equations and Formulas from Words

Take your time, read through the questions and extract all the important maths from them. It'll make it much easier to put into an equation or formula.

Q1　　Martin's mobile phone has a monthly charge of £15 and a charge of 8p per text message.

　　a)　Write a formula to work out how much in pounds Martin has to pay each month.

　　b)　Use your formula to calculate the bill if he sends 20 texts per month.

　　c)　Use your formula to calculate the bill if he sends 120 texts per month.

　　d)　Use your formula to calculate the bill if he sends 15 texts per week. Assume a month has 4 weeks.

> Hint: Call the number of texts n

Honestly, mobiles are perfectly safe...

Q2　　Paul's phone has no monthly charge but costs 50p for each daytime call and 25p for each evening call.

　　a)　Write a formula to calculate Paul's phone bill in pounds.

　　b)　Use the formula to work out his bill if he makes 10 daytime and 30 evening calls.

　　c)　Use the formula to work out his bill if he makes 5 daytime and 100 evening calls.

　　d)　If he made 4 daytime calls and his bill is £25, how many evening calls did he make?

　　e)　If he made 120 evening calls and his bill is £34, how many daytime calls did he make?

Q3　　At Jane's school there are 22 more girls than boys.

　　a)　Write an expression for the number of girls in terms of the number of boys, b.

　　　　　There are 460 pupils at the school.

　　b)　Using part a), write an equation involving b and solve it to find the number of boys at the school.

　　c)　How many girls are there at the school?

Q4　　The angles of a quadrilateral add up to 360°. Form an equation in x and solve it for each of the following shapes:

a)　　　　　　　**b)**　　　　　　　**c)**　　　　　　　**d)**

a) quadrilateral with angles $2x^0$, 90^0, x^0+34^0, x^0

b) quadrilateral with angles $x+8^0$, $x+7^0$, x^0, 90^0

c) quadrilateral with angles $3x^0$, $4.5x^0$, $3(x+5^0)$, x^0

d) quadrilateral with angles x^0, $6x+10^0$, $2x^0$, x^0

Rearranging Formulas

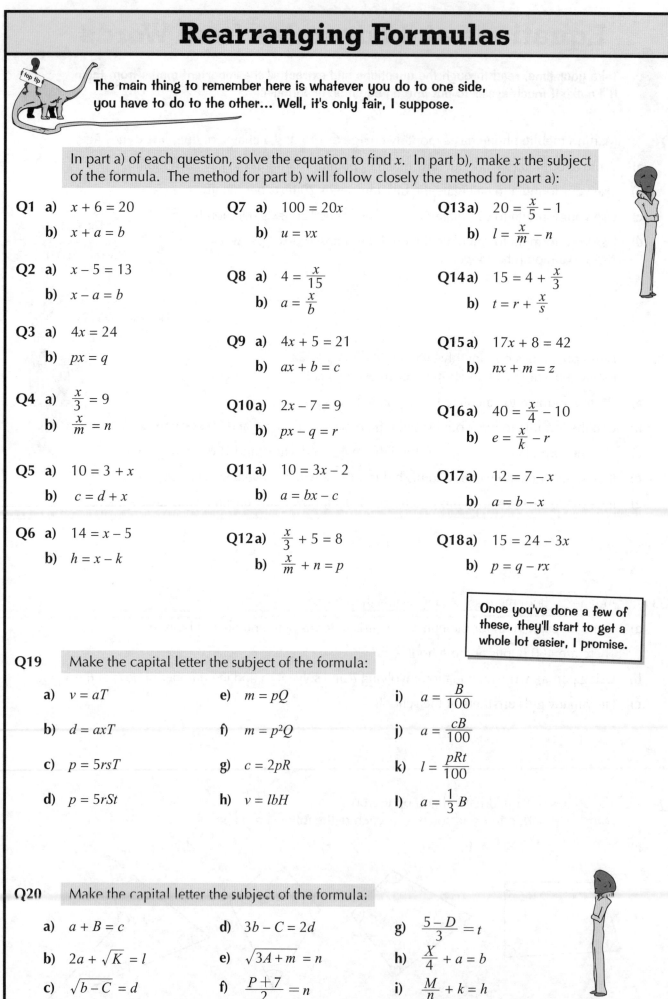

top tip

The main thing to remember here is whatever you do to one side, you have to do to the other... Well, it's only fair, I suppose.

In part a) of each question, solve the equation to find x. In part b), make x the subject of the formula. The method for part b) will follow closely the method for part a):

Q1 a) $x + 6 = 20$

b) $x + a = b$

Q2 a) $x - 5 = 13$

b) $x - a = b$

Q3 a) $4x = 24$

b) $px = q$

Q4 a) $\frac{x}{3} = 9$

b) $\frac{x}{m} = n$

Q5 a) $10 = 3 + x$

b) $c = d + x$

Q6 a) $14 = x - 5$

b) $h = x - k$

Q7 a) $100 = 20x$

b) $u = vx$

Q8 a) $4 = \frac{x}{15}$

b) $a = \frac{x}{b}$

Q9 a) $4x + 5 = 21$

b) $ax + b = c$

Q10 a) $2x - 7 = 9$

b) $px - q = r$

Q11 a) $10 = 3x - 2$

b) $a = bx - c$

Q12 a) $\frac{x}{3} + 5 = 8$

b) $\frac{x}{m} + n = p$

Q13 a) $20 = \frac{x}{5} - 1$

b) $l = \frac{x}{m} - n$

Q14 a) $15 = 4 + \frac{x}{3}$

b) $t = r + \frac{x}{s}$

Q15 a) $17x + 8 = 42$

b) $nx + m = z$

Q16 a) $40 = \frac{x}{4} - 10$

b) $e = \frac{x}{k} - r$

Q17 a) $12 = 7 - x$

b) $a = b - x$

Q18 a) $15 = 24 - 3x$

b) $p = q - rx$

Once you've done a few of these, they'll start to get a whole lot easier, I promise.

Q19 Make the capital letter the subject of the formula:

a) $v = aT$

b) $d = axT$

c) $p = 5rsT$

d) $p = 5rSt$

e) $m = pQ$

f) $m = p^2Q$

g) $c = 2pR$

h) $v = lbH$

i) $a = \frac{B}{100}$

j) $a = \frac{cB}{100}$

k) $l = \frac{pRt}{100}$

l) $a = \frac{1}{3}B$

Q20 Make the capital letter the subject of the formula:

a) $a + B = c$

b) $2a + \sqrt{K} = l$

c) $\sqrt{b - C} = d$

d) $3b - C = 2d$

e) $\sqrt{3A + m} = n$

f) $\frac{P + 7}{2} = n$

g) $\frac{5 - D}{3} = t$

h) $\frac{X}{4} + a = b$

i) $\frac{M}{n} + k = h$

Number Patterns and Sequences

I heard that number patterns and sequences are your favourite things, so here's a load of questions for you. Don't forget — finding the rule for a number pattern boils down to working out what you need to do to get from one number to the next.

Q1 Draw the next two pictures for each pattern. How many matchsticks are used in each picture?

a)

b)

c)

Q2 In this sequence the rule for getting each term is "Double n and add 1". Copy and complete the table containing the first 8 terms of the sequence.

n	1	2	3	4	5	6	7	8
t	3	5						

Q3 Describe in words the rule for finding the next term of each sequence and state whether the sequence is arithmetic or geometric. Then write down the next 3 terms.

a) 12, 14, 16

b) 224, 112, 56

c) 3, 6, 9, 12

d) 3, 6, 12, 24

e) 8, 6, 4, 2

f) -13, -10, -7, -4

g) -5, -8, -11, -14

h) 2, 6, 18, 54

Q4 In this sequence the rule for getting t is "Multiply n by 3, then subtract 2". Copy and complete the table up to the 8th term.

n	1	2	3	4	5	6	7	8
t								

Number Patterns and Sequences

This is a classic — you could easily get asked to find the nth term, so make sure you know how to do it. Always start by working out the difference between terms, then add or subtract to make your formula.

Q5   6   11   16   21   26 ...

a)   Copy this sequence and continue it for 3 more terms.

b)   Write the difference row underneath.

c)   Write a formula for calculating the term (t) in relation to the number of the term (n).

d)   What is the 20th term?

Q6   7   9   11   13   15 ...

a)   Copy this sequence and work out the next 3 terms.

b)   Write the difference row underneath.

c)   What is the nth term?

Q7   2   9   16   23   30 ...

a)   Copy the sequence and find the next 3 terms.

b)   Write a formula for the nth term.

c)   Is 107 a term in the sequence?

> To check if a value's in a sequence, just stick it into your formula. If n comes out as a whole number then the value is in the sequence.

Q8   4   6.5   9   11.5   14 ...

a)   Copy the sequence and continue it for 3 more terms.

b)   Work out a formula for the nth term.

c)   What is the 50th term?

Q9   20   17   14   11   8 ...

a)   Continue the sequence for 5 more terms.

b)   Write down the difference between each pair of consecutive terms.

c)   Write down a formula for the nth term.

d)   Is -78 a term in the sequence?

Q10   Find the nth term in each of the following sequences:

a)   3   10   17   24   31...

b)   5   9   13   17   21...

c)   14   11   8   5   2...

d)   27   22   17   12   7...

I said "another piggin' sequence" — not "a pig in sequins"...

Inequalities

Inequalities are a handy way of showing a range of values. Make sure you know what the symbols >, <, ≤ and ≥ mean — they're pretty important. Oh the joys of maths...

Q1 Write down all the possible values of x in the following inequalities.

a) x is a positive integer such that $x \leq 2$

b) x is a negative integer such that $x \geq -6$

c) x is a positive integer such that $x < 5$

d) x is a negative integer such that $x > -4$

Come 'ere Cheeky.

Q2 If n is an integer, write down all the values of n that satisfy the following inequalities.

a) $2 < n < 7$

b) $2 \leq n \leq 7$

c) $2 < n \leq 7$

d) $21 \leq n \leq 25$

e) $8 < n < 9.5$

f) $-2 < n < 4$

g) $-3 < n < 2$

h) $-7 \leq n \leq -3$

Q3 Show on a number line all values of n that obey these inequalities.

a) $n < 4$

b) $n > 1$

c) $4 < n \leq 8$

d) $0 \leq n \leq 6$

e) $-1 < n \leq 5$

f) $0.1 < n < 1.7$

g) $-3.5 < n < 1.5$

h) $1234 < n \leq 1237$

> Remember — use an open circle for < and a filled-in one for ≤.

Q4 Express using inequality signs, the range of values for each of the variables a, b, c, d, e, shown on this number line:

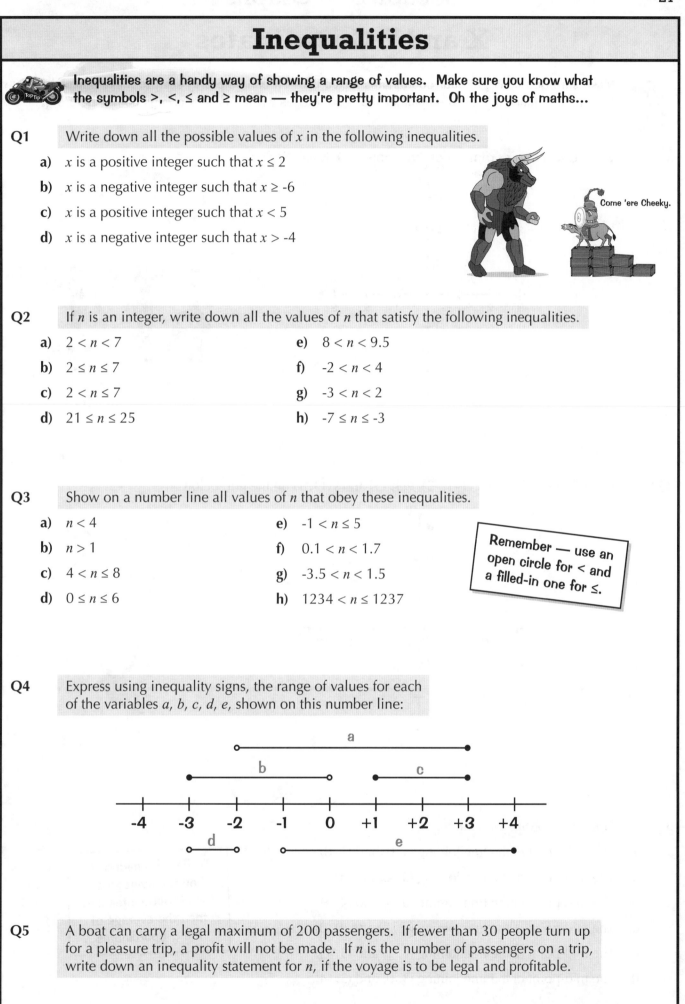

Q5 A boat can carry a legal maximum of 200 passengers. If fewer than 30 people turn up for a pleasure trip, a profit will not be made. If n is the number of passengers on a trip, write down an inequality statement for n, if the voyage is to be legal and profitable.

Section 3 — Graphs

X and Y Coordinates

Make sure you don't get your coordinates mixed up — it can be pretty easy to make a mistake. Your best defence is just to remember the order they're written in — x always comes before y.

Q1 Write down the coordinates of the points P, Q, R and S.

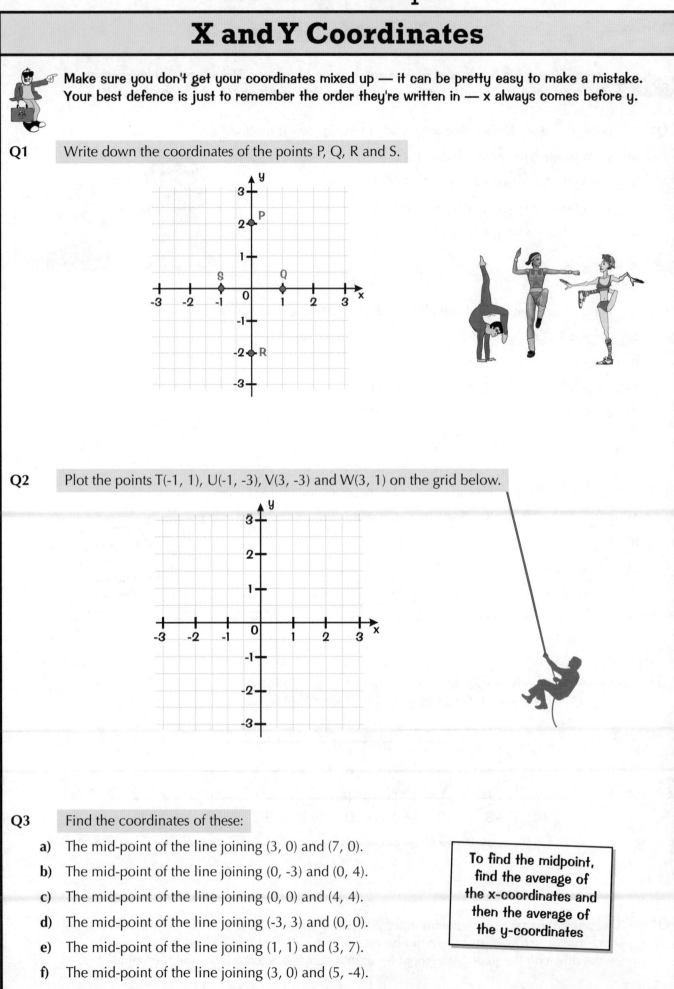

Q2 Plot the points T(-1, 1), U(-1, -3), V(3, -3) and W(3, 1) on the grid below.

Q3 Find the coordinates of these:

a) The mid-point of the line joining (3, 0) and (7, 0).

b) The mid-point of the line joining (0, -3) and (0, 4).

c) The mid-point of the line joining (0, 0) and (4, 4).

d) The mid-point of the line joining (-3, 3) and (0, 0).

e) The mid-point of the line joining (1, 1) and (3, 7).

f) The mid-point of the line joining (3, 0) and (5, -4).

> To find the midpoint, find the average of the x-coordinates and then the average of the y-coordinates

Plotting Straight Line Graphs

Got those coordinates sorted? Make sure you're happy with them before you plough headlong into a graph question. If not, it might go horribly wrong...

Q1   Draw axes from –5 to +5 in each direction, then plot the following three lines:

a) $y = 3x - 4$

b) $y = -\frac{4}{3}x - 1$

c) $y = \frac{1}{4}x + 1$

Start with the Table of Values and find 3 points on the line — then plot them on your graph and draw a line through them with your ruler.

Q2   Follow the instructions below:

a)   Draw x and y axes from –6 to +6.

b)   Rearrange the equations for the following four lines, into the form $y = mx + c$:

   i) $x + y = 4$

   ii) $3x + y = -6$

   iii) $x - 2y = -4$

   iv) $5x - 3y = -15$

c)   Plot the graphs of the four lines on your axes.

Q3   Draw x and y axes from -6 to +6.

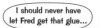

I should never have let Fred get that glue...

a)   Plot on your axes the 3 lines $y = 2x$, $y = 2x + 3$ and $y = 2x - 5$.

b)   On the same diagram, plot the lines $y = -2x$, $y = -2x + 3$ and $y = -2x - 5$.

c)   What do you notice about the pairs of lines:

   i) $y = 2x$ and $y = -2x$?

   ii) $y = 2x + 3$ and $y = -2x + 3$?

   iii) $y = 2x - 5$ and $y = -2x - 5$?

Q4   Draw axes from -6 to +6 in each direction.

a)   Plot the two lines $y = \frac{1}{2}x + 1$ and $y = -3x + 4\frac{1}{2}$.

b)   **i)**   Where does the first line cut the x-axis?

   ii)   Where does the second line cut the x-axis?

   iii)   Where do these lines cross each other?

Gradients and y = mx + c

Remember, y = mx + c. c is where it crosses the y-axis, and
m is the gradient. Practise, practise, practise...

Q1   Draw a graph with axes numbered from –6 to +6 in each direction.

a)   Plot the points from the tables below, and join them with straight lines.

x	-2	-1	0	1	2
y	-6	-3	0	3	6

x	-6	-3	0	3	6
y	-2	-1	0	1	2

b)   Use the graphs to work out the gradient of each line.
Show clearly on your graph how each gradient is worked out.

Best plan is to use points from the top-right quadrant to work out
your gradients, so that both x and y are positive. If they're not,
those dreaded minus signs could cause you real grief, so watch out.

Barbados

Q2   Draw another graph with axes numbered from –4 to +8 each way.

a)   Plot the points from the tables below and join them up to form two straight lines.

x	-1	0	1	2
y	-4	0	4	8

x	-4	0	4	8
y	-1	0	1	2

b)   On each line draw a triangle which could be used to find the gradient.

c)   Using your gradient triangles, determine the gradients of both of the lines.

Q3   Without drawing the graphs, determine the gradient of each of the following:

a) $y = 2x$

b) $y = 2x - 5$

c) $y = \frac{1}{4}x$

d) $y = \frac{1}{4}x + 3$

e) $y = -x$

f) $y = -x + 4$

g) $y = -\frac{1}{2}x$

h) $y = -\frac{1}{2}x - 1$

i) $y = -2x - 7$

j) $y = 5x - 7$

Q4   By first rearranging each equation into the form $y = mx + c$, find the gradient:

a) $x + 2y = 4$

b) $2x + 3y = 12$

c) $-x + 3y = 9$

d) $-3x + 4y = 12$

e) $x - 2y = 6$

f) $5x - 4y = 20$

g) $3x - 5y = -15$

h) $2x - 6y = -24$

Gradients and y = mx + c

Q5 Match the following straight line graphs with their equations:

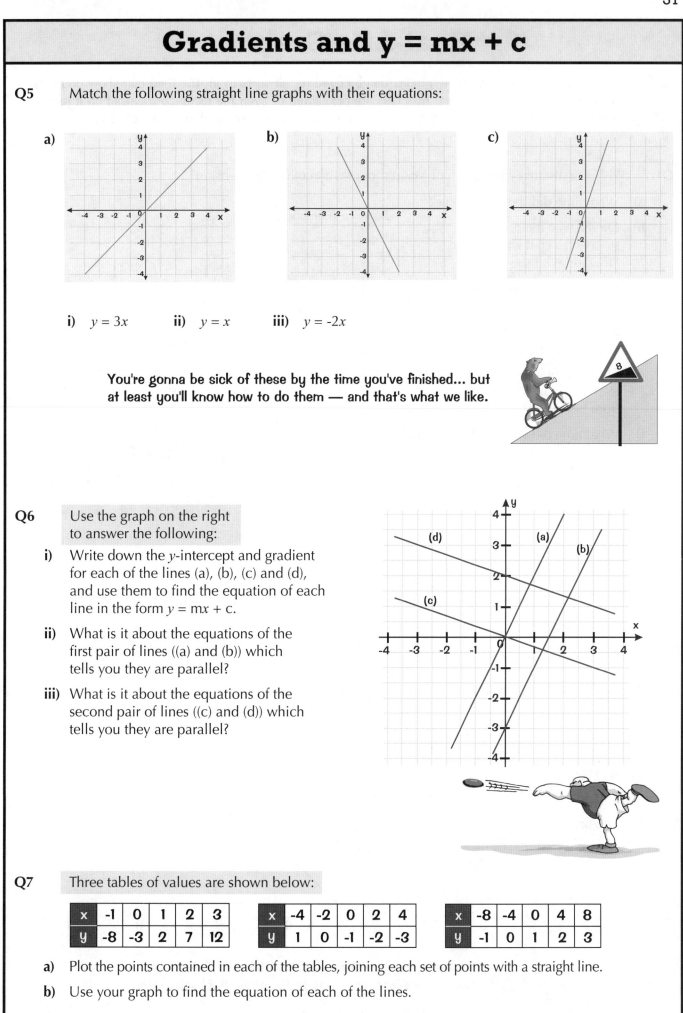

a)

b)

c)

i) $y = 3x$ ii) $y = x$ iii) $y = -2x$

You're gonna be sick of these by the time you've finished... but at least you'll know how to do them — and that's what we like.

Q6 Use the graph on the right to answer the following:

i) Write down the y-intercept and gradient for each of the lines (a), (b), (c) and (d), and use them to find the equation of each line in the form $y = mx + c$.

ii) What is it about the equations of the first pair of lines ((a) and (b)) which tells you they are parallel?

iii) What is it about the equations of the second pair of lines ((c) and (d)) which tells you they are parallel?

Q7 Three tables of values are shown below:

x	-1	0	1	2	3
y	-8	-3	2	7	12

x	-4	-2	0	2	4
y	1	0	-1	-2	-3

x	-8	-4	0	4	8
y	-1	0	1	2	3

a) Plot the points contained in each of the tables, joining each set of points with a straight line.

b) Use your graph to find the equation of each of the lines.

Real-Life Graphs

It's not all about x's and y's — graphs can be used for all sorts of stuff. They can show you how to convert between currencies, how bills are calculated or even the distance travelled on a journey. Make sure you pay attention to the axes so that you know exactly what the graph is showing.

Q1 This is a distance-time graph of Maud's journey to collect her pension.

a) What time did she leave home?

b) On the way there, she stopped to chat with Eric. How long did she stop for?

c) How far was the post office from her house?

d) How long did she have to wait in the queue at the post office?

e) On her way home she stopped to shop. How far was the shop from her house?

f) What time did she finally get home?

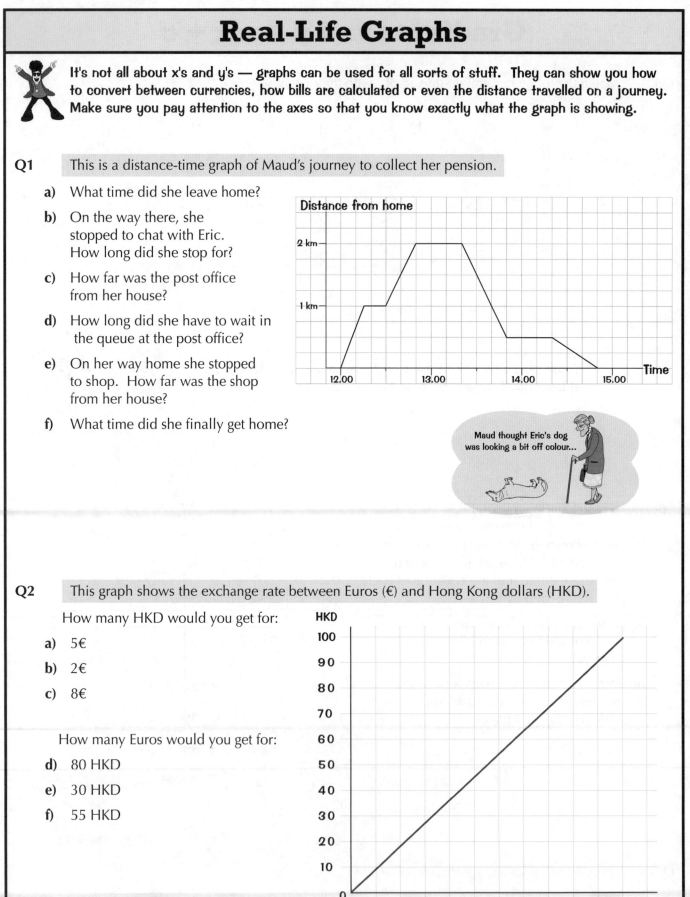

Maud thought Eric's dog was looking a bit off colour...

Q2 This graph shows the exchange rate between Euros (€) and Hong Kong dollars (HKD).

How many HKD would you get for:

a) 5€

b) 2€

c) 8€

How many Euros would you get for:

d) 80 HKD

e) 30 HKD

f) 55 HKD

Remember, conversion graphs can be read 2 ways — you can convert from one thing to the other and back again.

Real-Life Graphs

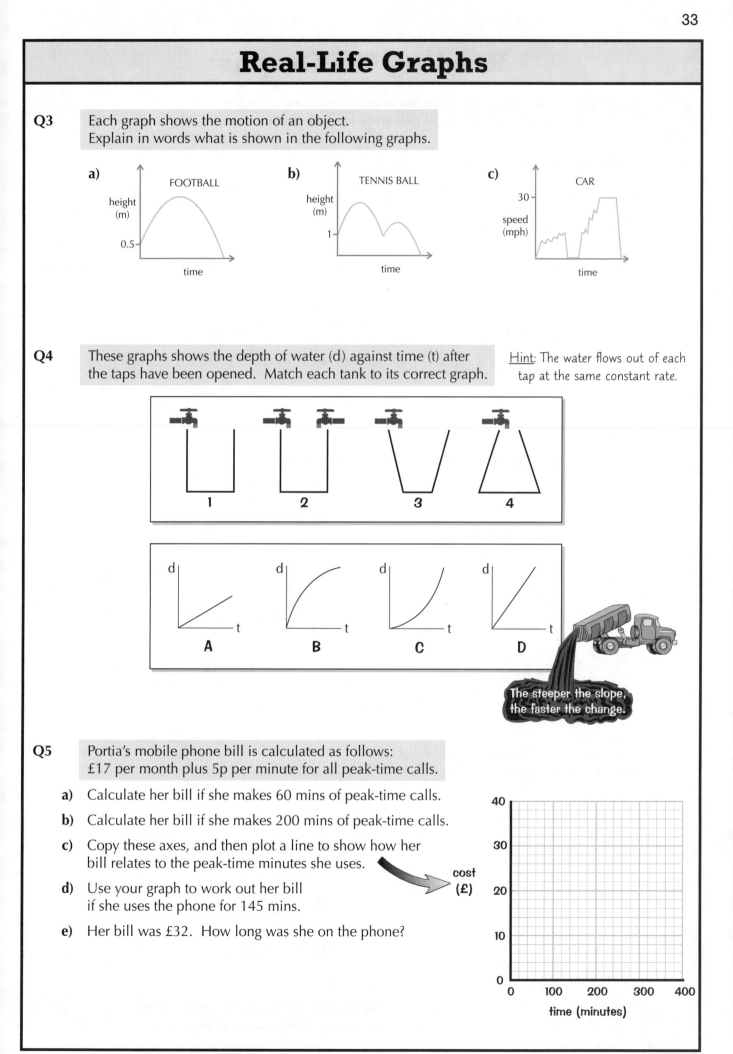

Q3   Each graph shows the motion of an object.
Explain in words what is shown in the following graphs.

a) FOOTBALL — height (m), 0.5, time

b) TENNIS BALL — height (m), 1, time

c) CAR — speed (mph), 30, time

Q4   These graphs shows the depth of water (d) against time (t) after
the taps have been opened. Match each tank to its correct graph.

Hint: The water flows out of each
tap at the same constant rate.

Tanks: 1, 2, 3, 4

Graphs: A, B, C, D

The steeper the slope,
the faster the change.

Q5   Portia's mobile phone bill is calculated as follows:
£17 per month plus 5p per minute for all peak-time calls.

a) Calculate her bill if she makes 60 mins of peak-time calls.

b) Calculate her bill if she makes 200 mins of peak-time calls.

c) Copy these axes, and then plot a line to show how her
bill relates to the peak-time minutes she uses.

d) Use your graph to work out her bill
if she uses the phone for 145 mins.

e) Her bill was £32. How long was she on the phone?

cost (£)

time (minutes)

34

Reading Off Graphs

Eyup, even more graphs. Here you get to solve equations using them —
it's a useful trick and might just get you some lovely marks in a test.

Q1 Using the graphs below, write down the following:

a)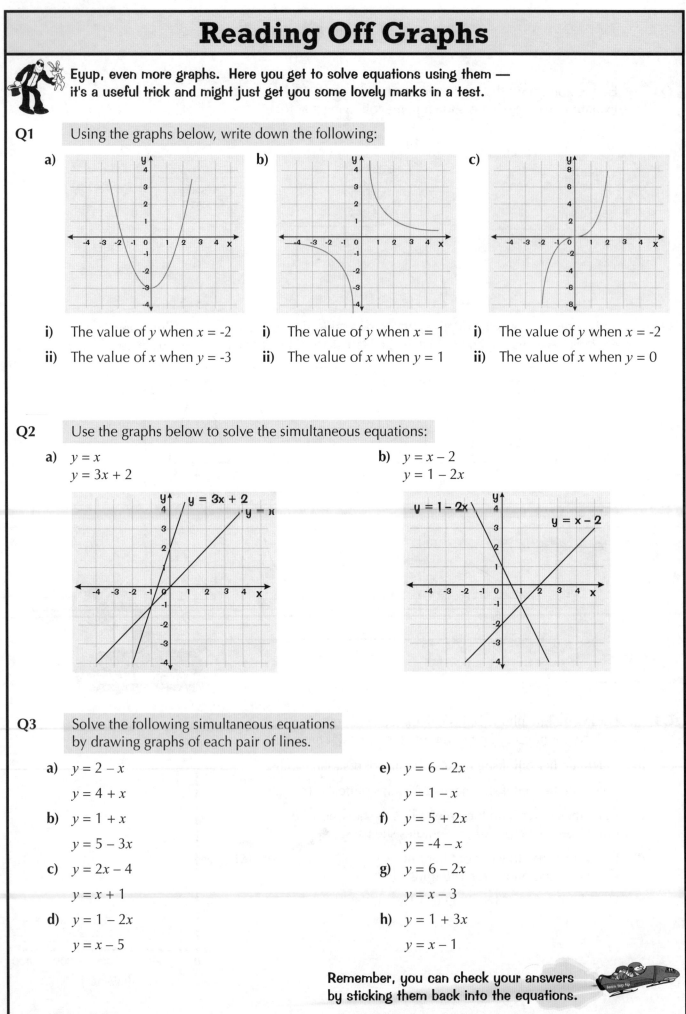

i) The value of y when $x = -2$

ii) The value of x when $y = -3$

b)

i) The value of y when $x = 1$

ii) The value of x when $y = 1$

c)

i) The value of y when $x = -2$

ii) The value of x when $y = 0$

Q2 Use the graphs below to solve the simultaneous equations:

a) $y = x$
$y = 3x + 2$

b) $y = x - 2$
$y = 1 - 2x$

Q3 Solve the following simultaneous equations by drawing graphs of each pair of lines.

a) $y = 2 - x$
$y = 4 + x$

b) $y = 1 + x$
$y = 5 - 3x$

c) $y = 2x - 4$
$y = x + 1$

d) $y = 1 - 2x$
$y = x - 5$

e) $y = 6 - 2x$
$y = 1 - x$

f) $y = 5 + 2x$
$y = -4 - x$

g) $y = 6 - 2x$
$y = x - 3$

h) $y = 1 + 3x$
$y = x - 1$

Remember, you can check your answers by sticking them back into the equations.

Section 3 — Graphs

Quadratic Graphs

Just when you thought graphs couldn't get any more complicated... sheesh.
Always check that the graph you've drawn is the right shape — remember that x^2 gives you a bucket shaped graph, and that if there's a '-' before the x^2 then the bucket will be upside down.

Q1 Copy and complete this table for $y = x^2$.

x	-4	-3	-2	-1	0	1	2	3	4	5
$y = x^2$	16							9		

a) Plot these points, using an x-axis from -4 to 5 and a y-axis from 0 to 25.

b) Join your points to make a smooth curve.

c) Use your curve to find y when $x = 3.5$.

d) What are the values of x when $y = 7$?

Q2 Copy and complete this table of values for the graph $y = x^2 - 2x + 2$.

x	-2	-1	0	1	2	3	4
x^2	4	1				9	
$-2x$	4					-6	
2	2	2				2	2
$x^2 - 2x + 2$	10	5				5	

a) Using your table of values, plot the graph $y = x^2 - 2x + 2$.

b) Draw and label the line of symmetry.

Q3 Copy and complete this table of values for the graph $4 - x^2$.

x	-3	-2	-1	0	1	2	3
4	4	4	4	4	4	4	4
$-x^2$	-9					-4	
$4 - x^2$	-5					0	

a) Using your table of values, draw the graph $y = 4 - x^2$.

b) Use your graph to estimate the solutions to the equation $4 - x^2 = 2$.

Q4 Draw the graph $y = x^2 - x - 2$ for values of x from -3 to 4.

Use your graph to estimate the solutions to the equation $x^2 - x - 2 = 6$.

Ratios

Welcome to another couple of pages of mind-blowing fun.
Ratios might look weird but they have a simple meaning —
if the ratio of cats to dogs is **2 : 3**, it means there are **2** cats for every **3** dogs.

Q1   Express the following ratios in their simplest form.

a)   15 : 20   e) $\frac{1}{4} : \frac{3}{4}$

b)   2 : 12   f)   3.2 : 0.8

c)   4 : 8 : 16   g)   0.2 : 1.2 : 0.6

d)   320 : 480   h)   0.02 : 0.2

Q2   Express these ratios in their simplest whole number form.

a)   12 teachers to 90 pupils.

b)   7 crew to 210 aircraft passengers.

c)   16 doctors to 32,000 people in town.

d)   8 guides to 120 tourists.

e)   350 g of flour to 150 g of sugar.

f)   30 cows to 180 turkeys.

> Giving ratios in their simplest form is a bit like cancelling down fractions... you can use the fraction button on your calculator to help too.

Q3   Write, in its simplest whole number form, the ratio of width to height for each rectangle in this diagram:

a)   Wall

b)   Door

c)   Window

d)   Picture

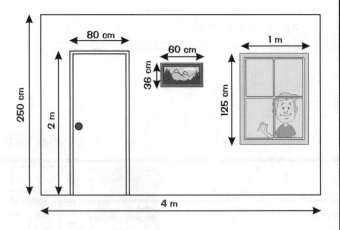

Q4   Express these ratios in the form 1 : n.

a)   2 : 3   e)   300 : 150

b)   5 : 8   f)   7 : 10

c)   8 : 5   g)   10 : 7

d) $\frac{3}{4} : \frac{3}{8}$ h)   6 : 1

Ratios

All we hear is, ratio ga ga, ratio goo goo... Never fear, only one more ratio page to go. For the "split something in the ratio..." questions, always check your answer by adding the bits back together — if you don't get the original amount, something's gone wrong.

Q5 A simple fruit salad is made, so that in each bowl there are 3 strawberries and 5 melon pieces. Copy and complete this chart:

No. of bowls	No. of strawberries	No. of melon pieces	Ratio of strawberries to melon pieces*
3			
	18		
		50	

* In its simplest whole number form

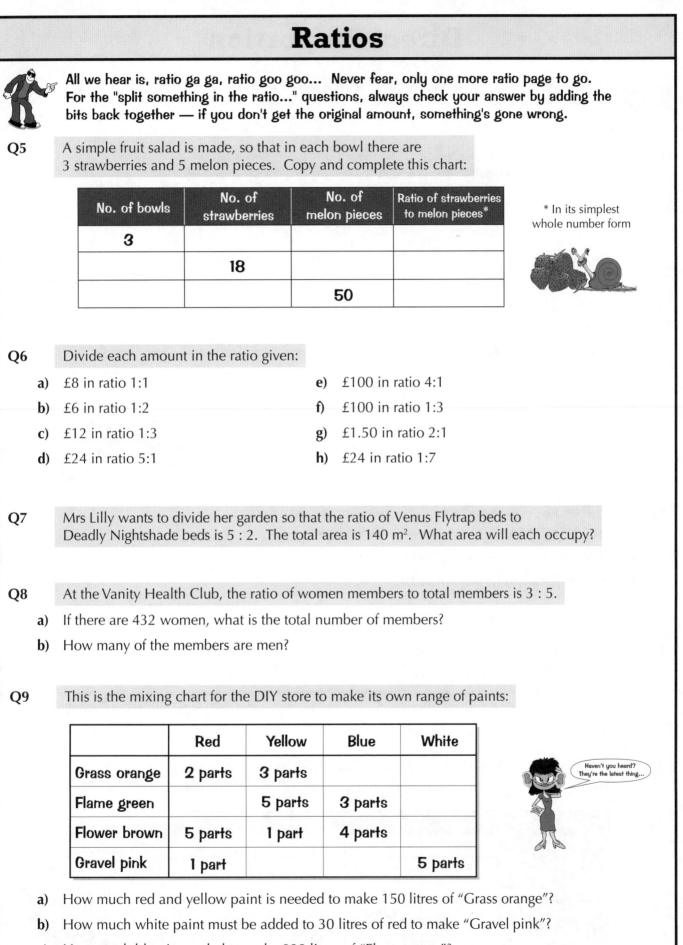

Q6 Divide each amount in the ratio given:

a) £8 in ratio 1:1

b) £6 in ratio 1:2

c) £12 in ratio 1:3

d) £24 in ratio 5:1

e) £100 in ratio 4:1

f) £100 in ratio 1:3

g) £1.50 in ratio 2:1

h) £24 in ratio 1:7

Q7 Mrs Lilly wants to divide her garden so that the ratio of Venus Flytrap beds to Deadly Nightshade beds is 5 : 2. The total area is 140 m². What area will each occupy?

Q8 At the Vanity Health Club, the ratio of women members to total members is 3 : 5.

a) If there are 432 women, what is the total number of members?

b) How many of the members are men?

Q9 This is the mixing chart for the DIY store to make its own range of paints:

	Red	Yellow	Blue	White
Grass orange	2 parts	3 parts		
Flame green		5 parts	3 parts	
Flower brown	5 parts	1 part	4 parts	
Gravel pink	1 part			5 parts

Haven't you heard? They're the latest thing...

a) How much red and yellow paint is needed to make 150 litres of "Grass orange"?

b) How much white paint must be added to 30 litres of red to make "Gravel pink"?

c) How much blue is needed to make 920 litres of "Flame green"?

d) If the DIY store is left with only 200 litres of blue paint, how much "Flower brown" paint can be made?

Direct Proportion

It may sound a bit confusing, but don't blow it out of proportion — it's just about dividing and multiplying. Remember, divide for <u>one</u>, then times for <u>all</u>.

Q1 Anthony is making cakes to sell at the school disco. Here is a list of ingredients for Chocolate Brownies. It makes enough for 20 brownies.

walnuts: 50g   eggs: 2   cocoa: 40g

butter: 50g   sugar: 225g   flour: 75g

a) How much of each ingredient will he need to make 120 brownies?

b) He finds he has only one 200 g tin of cocoa. How many brownies can he make?

Q2 A box of 25 chocolate bugs bars (with real bug centre) at the cash-and-carry costs £4.75.

a) How much does one bar cost?

b) How much do 6 bars cost?

c) Samson makes a 4p profit on every bar he resells. Write his profit as a fraction of the cost of each bar.

Q3 Tim just loves his perfect smile. He buys 5 m of dental floss from Supercheesygrin Dental Floss Supplies for £8.

a) How much would 6 m cost? (You pay by length.)

b) How much could he buy with £10?

Q4 A group of 80 angry bugs can eat 20 potatoes a day.

a) How many potatoes could 140 bugs eat a day?

b) Write an equation for this proportion in terms of potatoes (p) eaten and number of bugs (b).

Q5 A former popstar from the band Z-Club 6 is selling signed goodies at a local church fete. He makes a table to show how much money he will make. Copy and complete the table.

Amount Bought	Singles	Albums	T-Shirts	Plastic Dolls
1	25p			
2	50p			
3				
4		68p		
5			£1.10	
6				78p

Ee, they don't make em like Z-Club anymore...

Inverse Proportion

Apparently, my friends' happiness is inversely proportional to the amount of singing I do...
These sorts of questions can look very confusing, but there's nowt to worry about if you
remember the key rule — multiply for <u>one</u>, then divide for <u>all</u>.

Q1   It takes 2 badgers 3 minutes to eat a worm pie.
How long would it take 4 badgers to eat the same pie?

Q2   It took 3 grandparents 27 hours to program a new video
game. Assuming all grandparents work at the same
rate, how long would it have taken 9 grandparents?

Q3   It takes 150 minutes for 2 bakers to make 75 identical sausage rolls.
How long would it take 12 bakers to make the same number of sausage rolls?

Q4   Stanley goes for a run. If he runs at 6 miles per hour it will take 30 minutes.
How long will it take if he runs the same distance at 8 miles per hour?

Q5   It takes 2 mechanics 135 minutes to change 12 car tyres.

a)   How long would it take 9 mechanics to change 12 car tyres?

b)   Let x be the number of mechanics and t be the number of minutes it takes them to change
12 tyres. Write an equation in the form $t = \frac{A}{x}$ to represent this inverse proportion.

Just stick the numbers you know into
the equation, then rearrange to find A.

Q6   It takes 10 builders 3 months to build a bungalow.

a)   How long would it take 15 builders to build a bungalow?

b)   Let b be the number of builders and m be the number of months it takes them to build a bungalow.
Write an equation in the form $m = \frac{A}{b}$ to represent this inverse proportion.

Percentage Change

You may have already done a few percentage questions, but 'Percentage Change' is where all the cool kids hang out...

Q1   A restaurant bill comes to £46.00 for a family, then V.A.T. is added at 20%.

    a)   How much money is the V.A.T.?

    b)   How much does the family actually have to pay?

Q2   John has just had a 4% pay rise. He used to earn £238.50 a week. What will he now be earning?

You need to work out the total amount he'll be earning, not just the extra bit.

Q3   Last year Tumbledown Towers had 14350 visitors. This year they are expecting an increase of 18%. How many are they expecting altogether this year?

Q4   Glumchester Police report a 14% decrease in burglaries since last year, when 450 burglaries were recorded. How many have there been this year?

Q5   I put £350 into a building society savings account, which pays 6% simple interest each year. I didn't pay in or take out any other money, and after 3 years I went to check how much interest I had been paid.

    a)   Calculate the amount of interest I should have been paid each year.

    b)   Determine how much money I had in the account at the end of the 3 years.

Q6   When Shane bought a sports car for £7995, he expected it to depreciate (lose value) by 30% over two years. In fact, he sold it after two years for £5950.

    a)   What was his actual percentage loss (to the nearest whole number)?

    b)   To the nearest £10, how much better off was he than expected?

Q7   Glen bought a house for £75,000 and later sold it for £85,500.

    a)   What was the percentage gain in value?

    b)   In fact he had to pay an agent £1495, a solicitor £650, and a removal company £750, in order to move house. What was his actual percentage gain after these deductions? (1 decimal place)

Percentage Change

Don't let all the words put you off — just pick out the numbers and look for whether the percentage change is an increase or decrease.

Q8 After a 6% rise, Monty is getting £265 a week. What was he getting before? (£265 must be 106% of the original amount.)

Q9 A telephone bill comes to £49.28 after a 12% deduction is made. What was it originally? (£49.28 must be 88% of the original amount.)

Calculations — must have more calculations.

Q10 Washout Water Company has a problem with leaking pipes. 504 000 litres reach customers daily, but this is 28% less than the amount of water that enters the pipes. How much water is really being used?

Q11 The Jones family paid £1890 for their holiday with Shark Tours after a 12.5% surcharge was added at the last minute. What did they originally think they would be paying?

Q12 The Atlantic Ocean covers about 94 million square kilometres, which is about 26% of the total area of the world's oceans. What is the total area of the world's oceans? Give your answer to 1 d.p.

Q13 The Old Fleece Inn billed Mr McTavish £24.15 for a 3 course meal. This included a 15% service charge, which Mr McTavish refused to pay because the waiter had spilt gravy on his best suit. The manager withdrew the service charge — what did Mr McTavish actually pay in the end?

Quack

Q14 Gary has just run the 100 metres in 9.9 seconds, which was a 10% improvement on his previous best time. What was that best time?

Converting Units

Converting units can sound a little hairy, but chill man — it's mainly a matter of knowing the conversion factors. Once you've written down the correct one, decide whether to multiply or divide by it — that'll depend on whether you expect your answer to come out bigger or smaller than the original number.

Q1 Convert these measurements from centimetres into millimetres:

a) 4.8 cm   c) 8.75 cm

b) 26.4 cm   d) 0.63 cm.

Q2 Convert these measurements from millimetres into centimetres:

a) 76 mm   c) 3500 mm

b) 185 mm   d) 0.5 mm.

Q3 Rewrite these lengths in metres:

a) 145 cm   d) 5 cm

b) 350 cm   e) 2500 cm

c) 85 cm   f) 15.5 cm.

Q4 Convert these lengths into centimetres: 5 m, 5.6 m, 5.68 m, 0.75 m, 0.05 m.

Q5 Rewrite these weights entirely in grams: 1.4 kg, 2.85 kg, 0.65 kg

Q6 Convert these weights into kilograms: 450 g, 1450 g, 2450 g, 50 g, 5 g.

Q7 A large jug holds 3.0 litres of blood.
How many 150 ml glasses can be filled from it?

Q8 The average weight of the passengers on a bus is 58 kg each.
If the bus is full up with 71 passengers, what is their total weight in tonnes?

Q9 The water butt in my garden holds 15 gallons of rain-water. How many litres is this?

Q10 A flight from London to New York takes 7 hours and 17 minutes. How many minutes is this?

Q11 Giles walked 9 km in one day, while Bob walked 5 miles. Who walked further?

Converting Units

Q12   The table shows the distances in miles between 4 towns in Scotland.
Fill in the blank table with the equivalent distances in kilometres.

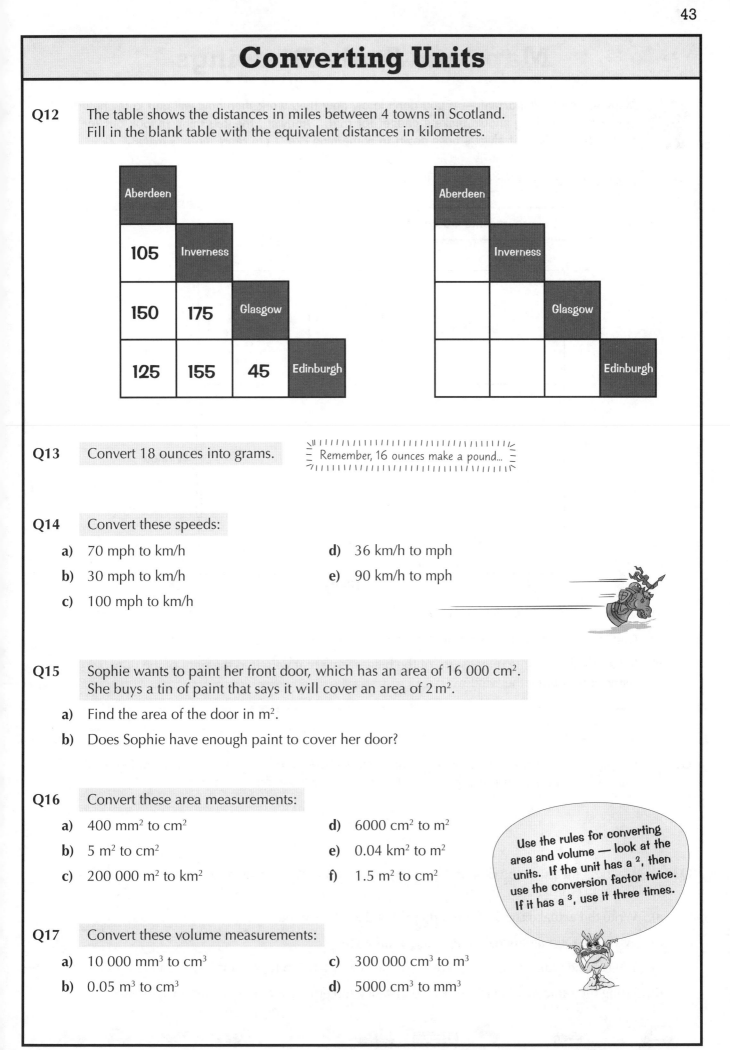

Q13   Convert 18 ounces into grams.

Remember, 16 ounces make a pound...

Q14   Convert these speeds:

a)   70 mph to km/h

b)   30 mph to km/h

c)   100 mph to km/h

d)   36 km/h to mph

e)   90 km/h to mph

Q15   Sophie wants to paint her front door, which has an area of 16 000 cm².
She buys a tin of paint that says it will cover an area of 2 m².

a)   Find the area of the door in m².

b)   Does Sophie have enough paint to cover her door?

Q16   Convert these area measurements:

a)   400 mm² to cm²

b)   5 m² to cm²

c)   200 000 m² to km²

d)   6000 cm² to m²

e)   0.04 km² to m²

f)   1.5 m² to cm²

Use the rules for converting area and volume — look at the units. If the unit has a ², then use the conversion factor twice. If it has a ³, use it three times.

Q17   Convert these volume measurements:

a)   10 000 mm³ to cm³

b)   0.05 m³ to cm³

c)   300 000 cm³ to m³

d)   5000 cm³ to mm³

Maps and Scale Drawings

Maps are pretty important — if you don't know how they work then how will you know how to get to places? They should make a gadget for that... Anyway, the key thing to look out for here is the scale — it always boils down to something like "1 cm represents 3 miles".

Q1 Here is a scale drawing of Ophelia's bedroom.

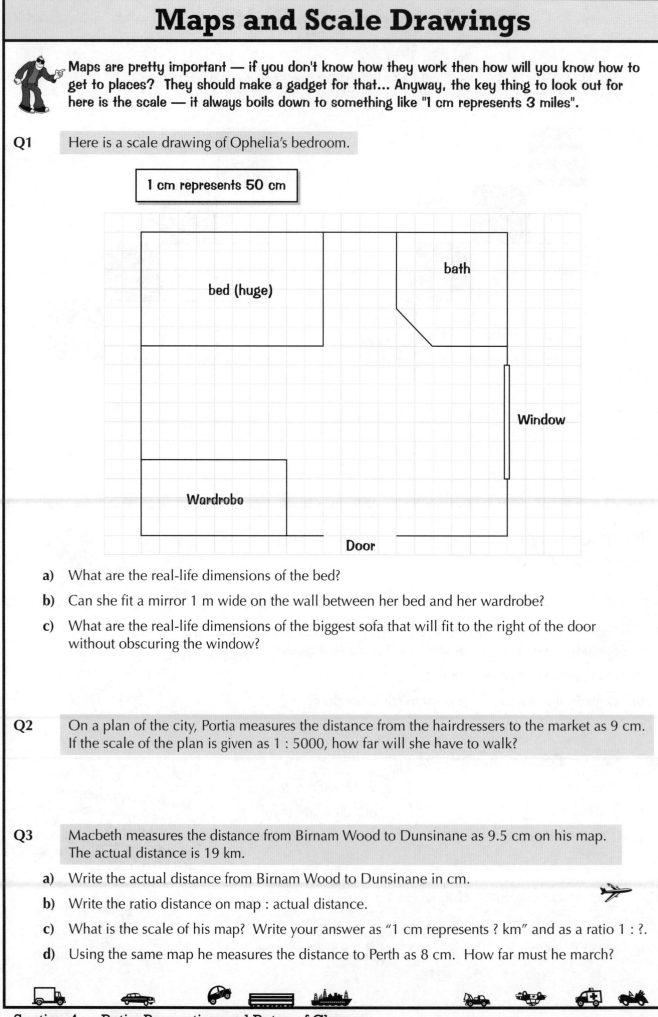

1 cm represents 50 cm

bed (huge)

bath

Window

Wardrobe

Door

a) What are the real-life dimensions of the bed?

b) Can she fit a mirror 1 m wide on the wall between her bed and her wardrobe?

c) What are the real-life dimensions of the biggest sofa that will fit to the right of the door without obscuring the window?

Q2 On a plan of the city, Portia measures the distance from the hairdressers to the market as 9 cm. If the scale of the plan is given as 1 : 5000, how far will she have to walk?

Q3 Macbeth measures the distance from Birnam Wood to Dunsinane as 9.5 cm on his map. The actual distance is 19 km.

a) Write the actual distance from Birnam Wood to Dunsinane in cm.

b) Write the ratio distance on map : actual distance.

c) What is the scale of his map? Write your answer as "1 cm represents ? km" and as a ratio 1 : ?.

d) Using the same map he measures the distance to Perth as 8 cm. How far must he march?

Best Buy

It's dead handy to be able to work out which item is the best value for money —
it'll definitely come in useful in real life, and it's also pretty likely to come up in a test.
So get it sorted and you'll bag yourself some marks and save loads of dosh. Brilliant.

Q1   I can either buy 6 m of ribbon for £2.50, or 7 m of ribbon for £2.60.
Which is the better buy?

Q2   My local bakery is doing a special batch of olive bread,
which will be sold in three different sized loaves.

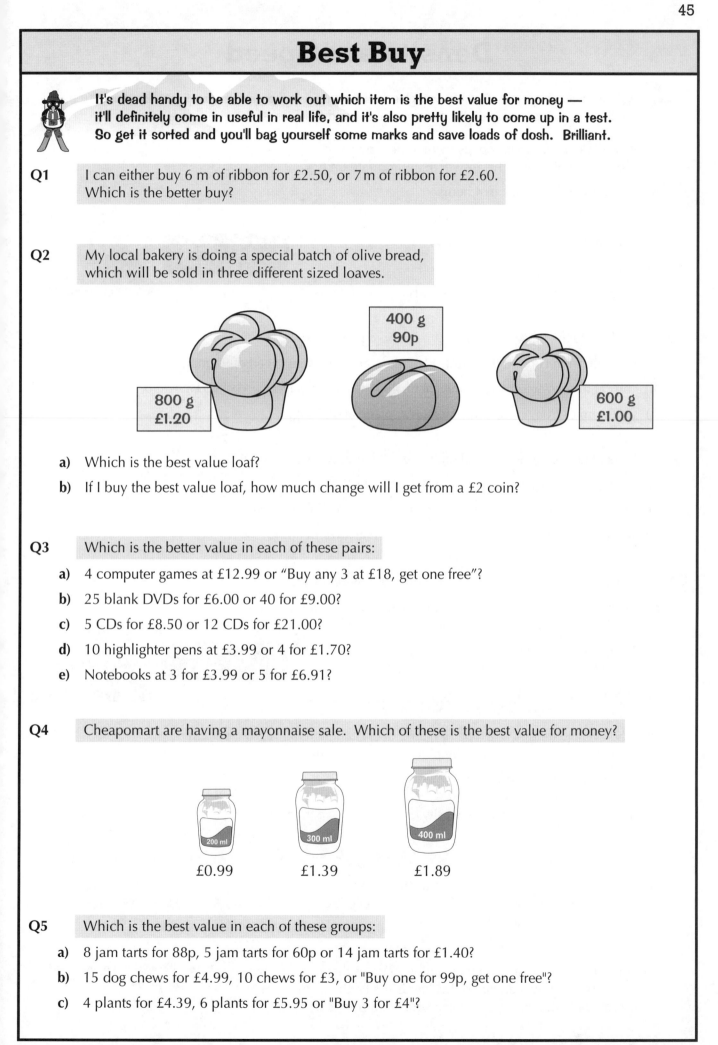

400 g
90p

800 g
£1.20

600 g
£1.00

   a)   Which is the best value loaf?

   b)   If I buy the best value loaf, how much change will I get from a £2 coin?

Q3   Which is the better value in each of these pairs:

   a)   4 computer games at £12.99 or "Buy any 3 at £18, get one free"?

   b)   25 blank DVDs for £6.00 or 40 for £9.00?

   c)   5 CDs for £8.50 or 12 CDs for £21.00?

   d)   10 highlighter pens at £3.99 or 4 for £1.70?

   e)   Notebooks at 3 for £3.99 or 5 for £6.91?

Q4   Cheapomart are having a mayonnaise sale. Which of these is the best value for money?

200 ml

300 ml

400 ml

£0.99   £1.39   £1.89

Q5   Which is the best value in each of these groups:

   a)   8 jam tarts for 88p, 5 jam tarts for 60p or 14 jam tarts for £1.40?

   b)   15 dog chews for £4.99, 10 chews for £3, or "Buy one for 99p, get one free"?

   c)   4 plants for £4.39, 6 plants for £5.95 or "Buy 3 for £4"?

Density and Speed

Whoever came up with formula triangles was a genius — they make questions like these really easy... just cover up the thing you want and the triangle tells you what to do.

Q1 Calculate the average speeds in the following cases, being careful to give the answer in appropriate units:

a) a car going 180 miles in 4 hours

b) a hiker walking 26 miles in 8 hours

c) a train going 725 km in 5.8 hours

d) a plane flying 3500 km in 5.6 hours

e) a rocket going 240 km in 30 seconds

$$\text{Speed} = \frac{\text{Distance}}{\text{Time}}$$

D
S × T

Q2 Calculate the time required for the following journeys:

a) by train from London to Bristol, 118 miles, at an average speed of 92 mph (2 d.p.)

b) by car from Worcester to Birmingham, 49 km, at an average speed of 60 km/h (2 d.p.)

c) 3500 miles from Europe to America by plane at 560 mph.

Q3 Calculate the distance travelled in the following cases:

a) a soggy pea moving at 25 m/s for 18.5 seconds

b) a snail moving at 0.3 cm/s for 2 minutes (give answer in metres.)

c) a river flowing at 4.5 m/s for 24 hours (give answer in km.)

Q4 Use the density formula triangle to answer the following:

a) A small block of silver has a volume of 4.5 cm³ and a mass of 47.25 g. Calculate the density of the silver.

b) Mercury has a density of 13.6 g per cm³. Calculate the mass of 6.5 cm³ of mercury.

$$\text{Density} = \frac{\text{Mass}}{\text{Volume}}$$

M
D × V

c) The density of gold is 19.3 g/cm³. Find the volume of a lump of gold with mass 57.9 g.

d) The average density of a type of quartz (a crystalline rock) is 2.6 g per cm³. What will be the mass of a lump of quartz of volume 64 cm³?

e) Ice has a density of about 0.9 g per cm³. What volume of ice will be produced when 810 ml of water freezes? (Take 1 ml = 1 cm³ and the density of water as 1 g per cm³.)

Q5 The mean density of the Earth is about 5.52 g per cm³. That of the Moon is only 3.34 g per cm³.

a) If the Earth had the same volume as the Moon, how many times heavier than the Moon would it be? Give your answer to 2 sf.

You haven't been given the volume, so just call it V and it'll cancel out later on. It will, trust me.

b) In fact, the Earth has a volume about 49 times that of the Moon. How many times heavier is the Earth compared with the Moon? Give your answer to 2 sf.

This time, call the volume of the Moon V and the volume of the Earth 49V... the V still cancels out nicely.

Symmetry

Fun fact: This page contains more lines of symmetry than
any other page in this book... What a treat.

Q1 These shapes have more than one line of symmetry.
Draw the lines of symmetry using dotted lines.

Make sure you've
got that ruler out...

a)

b)

c)

Q2 Write down the order of rotational symmetry of each of the following shapes:

a)

square

b)

rectangle

c)

equilateral
triangle

d)

parallelogram

> You can work out the rotational symmetry by sticking your pen in the
> middle of the shape and spinning your paper round — see how many
> times the shape looks the same before the paper's back the right way up.

Q3 For each of the capital letters below, write down: **a)** the number of lines of symmetry.
b) the order of rotational symmetry.

i)

N

ii)

T

iii)

S

iv)

C

Q4 Complete the following diagrams so that they have
rotational symmetry about centre C of the order stated:

a) order 2

b) order 4

c) order 3

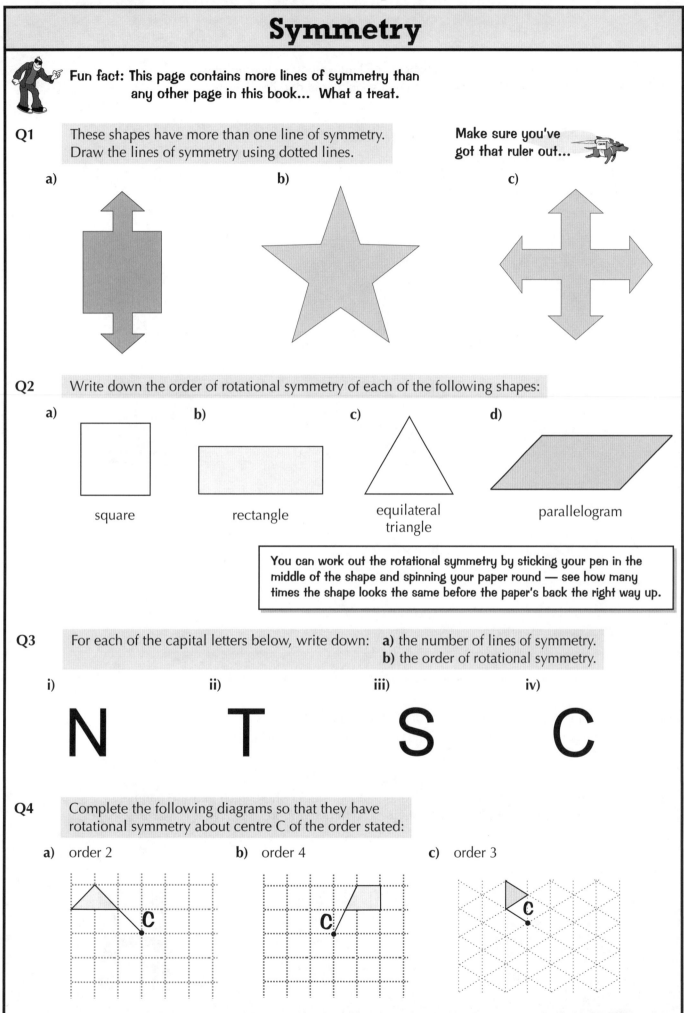

2D Shapes

Oooh, some more pretty shapes. As long as you've learned the properties of 2D shapes, this page should be a doddle. Lovely.

I'm just a regular kind of guy...

Q1 For each of the following shapes, write down:
i) The name of the shape.
ii) The number of lines of symmetry.
iii) The order of rotational symmetry.

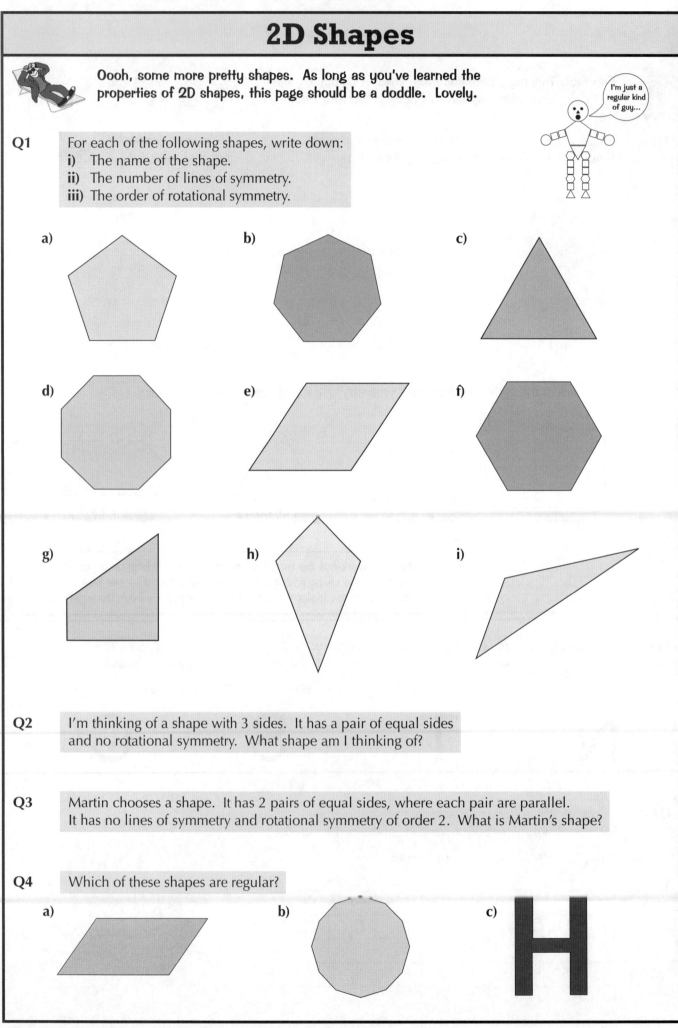

a)

b)

c)

d)

e)

f)

g)

h)

i)

Q2 I'm thinking of a shape with 3 sides. It has a pair of equal sides and no rotational symmetry. What shape am I thinking of?

Q3 Martin chooses a shape. It has 2 pairs of equal sides, where each pair are parallel. It has no lines of symmetry and rotational symmetry of order 2. What is Martin's shape?

Q4 Which of these shapes are regular?

a)

b)

c)

Perimeter and Area

Areas come up all the time in tests, so you need to get to grips with them.
Unfortunately, there's really only one way to win here, and that's to learn the formulas.

Q1   What is the area of a square field with sides 0.3 km long?

Easy marks... length times width, and there's your area. See — Maths is simple really.

Q2   Find the areas of these rectangles in square metres:

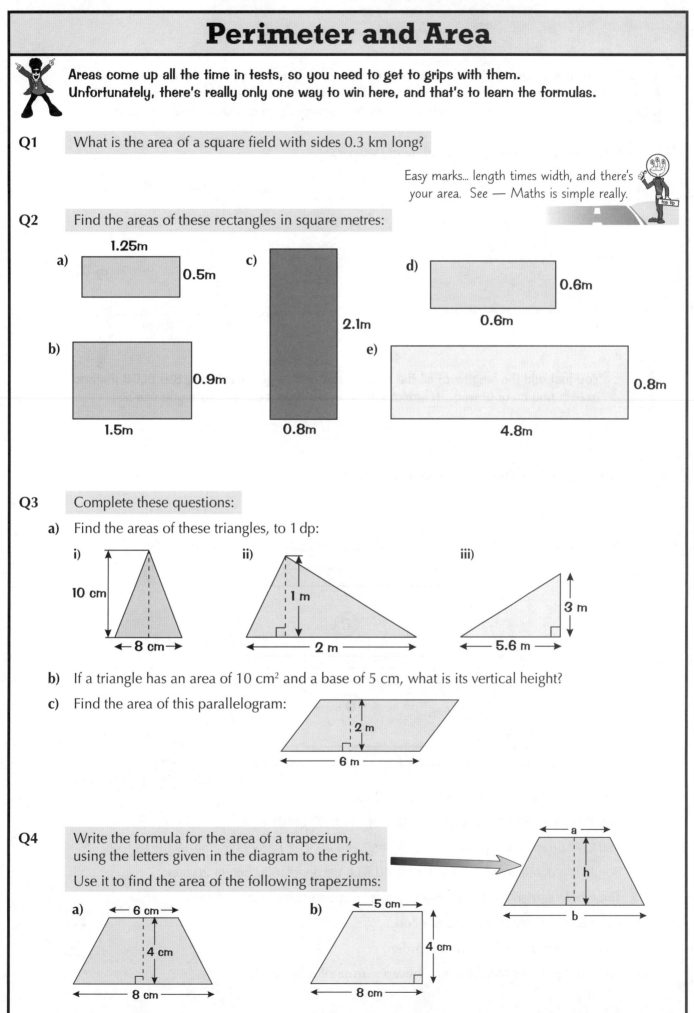

a)   1.25m   0.5m

b)   0.9m   1.5m

c)   2.1m   0.8m

d)   0.6m   0.6m

e)   0.8m   4.8m

Q3   Complete these questions:

a)   Find the areas of these triangles, to 1 dp:

i)   10 cm   ← 8 cm →

ii)   1 m   ← 2 m →

iii)   3 m   ← 5.6 m →

b)   If a triangle has an area of 10 cm² and a base of 5 cm, what is its vertical height?

c)   Find the area of this parallelogram: 2 m   ← 6 m →

Q4   Write the formula for the area of a trapezium, using the letters given in the diagram to the right.

Use it to find the area of the following trapeziums:

a)   ← 6 cm →   4 cm   ← 8 cm →

b)   ← 5 cm →   4 cm   ← 8 cm →

a   h   b

Perimeter and Area

Yep, there's more. Sorry. There's a bumper area question to look forward to at the end of the page — when you're finding the areas of complicated shapes, remember to split them into bits which you can work out separately, then add or subtract them.

Q5 Work out the perimeter of the following shapes:

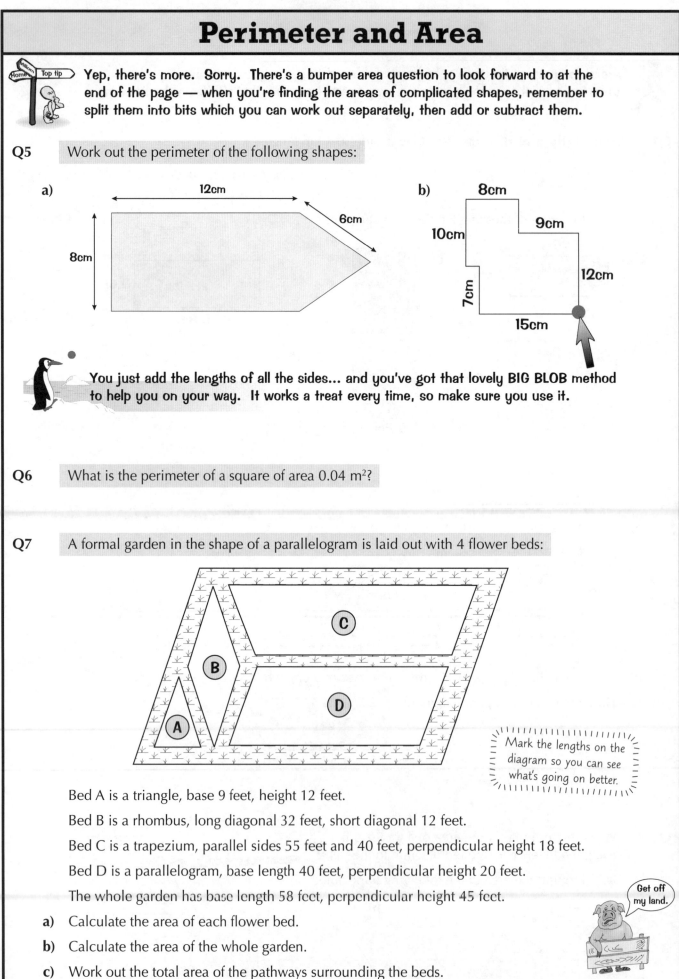

a)

12cm

6cm

8cm

b)

8cm

9cm

10cm

7cm

12cm

15cm

You just add the lengths of all the sides... and you've got that lovely **BIG BLOB** method to help you on your way. It works a treat every time, so make sure you use it.

Q6 What is the perimeter of a square of area 0.04 m²?

Q7 A formal garden in the shape of a parallelogram is laid out with 4 flower beds:

C

B

D

A

Mark the lengths on the diagram so you can see what's going on better.

Bed A is a triangle, base 9 feet, height 12 feet.

Bed B is a rhombus, long diagonal 32 feet, short diagonal 12 feet.

Bed C is a trapezium, parallel sides 55 feet and 40 feet, perpendicular height 18 feet.

Bed D is a parallelogram, base length 40 feet, perpendicular height 20 feet.

The whole garden has base length 58 feet, perpendicular height 45 feet.

Get off my land.

a) Calculate the area of each flower bed.

b) Calculate the area of the whole garden.

c) Work out the total area of the pathways surrounding the beds.

Circles

Q1   Calculate the circumference (to 1 dp) of the following.

  a) a wheel with diameter 65 cm

  b) a circular table with diameter 88 cm

  c) a can with diameter 8.5 cm

  d) a bowl with diameter 15.3 cm

  e) a circus ring with diameter 21 m

  f) the moon (diameter 2160 miles, answer to nearest 10 miles).

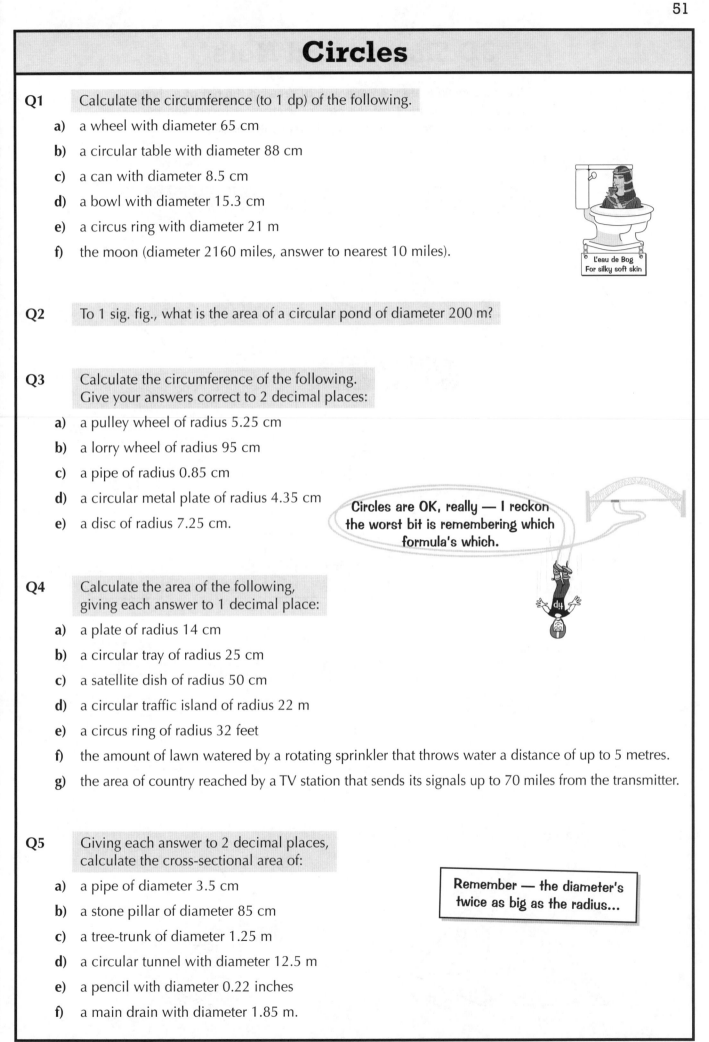

L'eau de Bog
For silky soft skin

Q2   To 1 sig. fig., what is the area of a circular pond of diameter 200 m?

Q3   Calculate the circumference of the following.
          Give your answers correct to 2 decimal places:

  a) a pulley wheel of radius 5.25 cm

  b) a lorry wheel of radius 95 cm

  c) a pipe of radius 0.85 cm

  d) a circular metal plate of radius 4.35 cm

  e) a disc of radius 7.25 cm.

Circles are OK, really — I reckon
the worst bit is remembering which
formula's which.

Q4   Calculate the area of the following,
          giving each answer to 1 decimal place:

  a) a plate of radius 14 cm

  b) a circular tray of radius 25 cm

  c) a satellite dish of radius 50 cm

  d) a circular traffic island of radius 22 m

  e) a circus ring of radius 32 feet

  f) the amount of lawn watered by a rotating sprinkler that throws water a distance of up to 5 metres.

  g) the area of country reached by a TV station that sends its signals up to 70 miles from the transmitter.

Q5   Giving each answer to 2 decimal places,
          calculate the cross-sectional area of:

  a) a pipe of diameter 3.5 cm

  b) a stone pillar of diameter 85 cm

  c) a tree-trunk of diameter 1.25 m

  d) a circular tunnel with diameter 12.5 m

  e) a pencil with diameter 0.22 inches

  f) a main drain with diameter 1.85 m.

Remember — the diameter's
twice as big as the radius...

3D Shapes and Nets

Don't forget that there's usually more than one way of drawing the net of a shape. That might come in handy. Also, don't get the nets of shapes confused with the other kind of nets — you won't catch many fish with these I'm afraid...

Q1 Match these three names with the 2-dimensional drawings of the 3-D shapes.

i) tetrahedron

ii) square based pyramid

a)

b)

c)

iii) triangular prism

Two of these names speak for themselves... and as for the tetrahedron, the "<u>tetra</u>" bit comes from the Greek for <u>4</u> and "<u>hedron</u>" kind of means "<u>faces</u>"... you work the rest out.

Q2 Which of the following nets would make a cube?

a)

b)

c)

d)

e)

f)

Q3 Draw an accurate net for each of the following solid shapes.

That means get your ruler out...

a)

5cm

3cm

4cm

b)

3cm

4cm

2cm

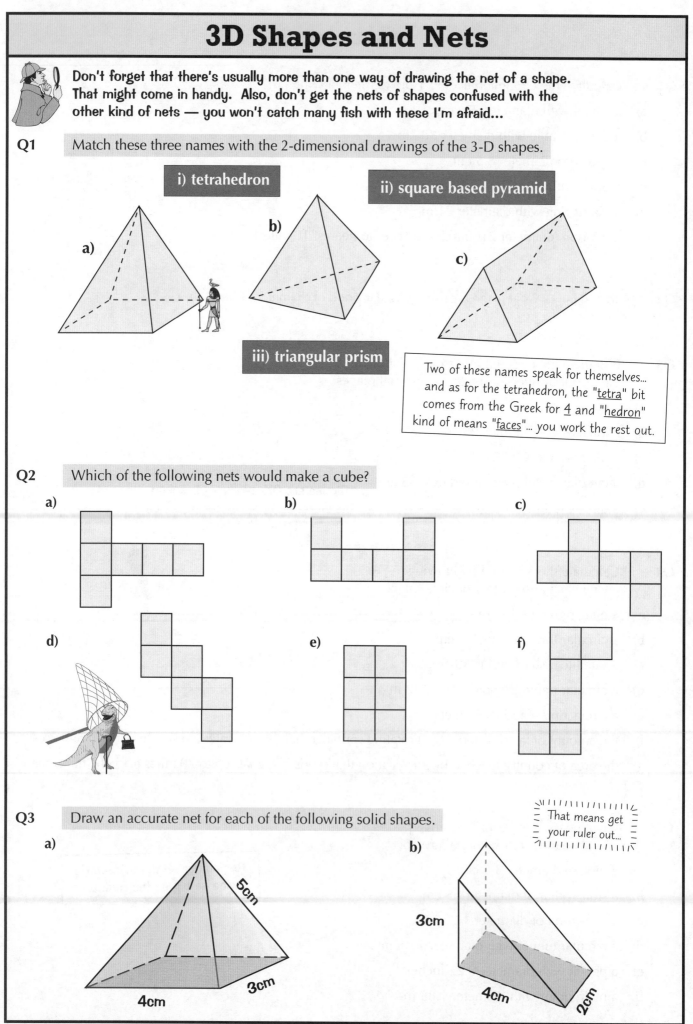

Surface Area

Surface area is the area of all the faces of a 3D shape added together.
Sometimes you'll find it's easier to work it out if you draw out a net first, but I'll not force you into it — that would be mean. Just bear it in mind if you're struggling.

Q1   Work out the surface area of each of the cuboids below:

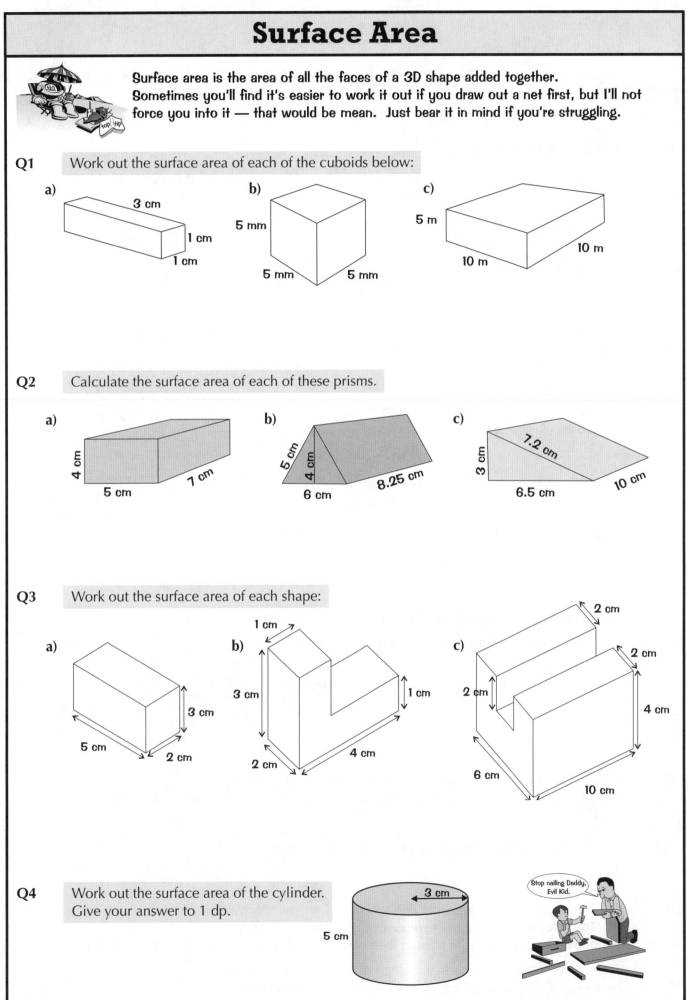

a)   3 cm   1 cm   1 cm

b)   5 mm   5 mm   5 mm

c)   5 m   10 m   10 m

Q2   Calculate the surface area of each of these prisms.

a)   4 cm   5 cm   7 cm

b)   5 cm   4 cm   6 cm   8.25 cm

c)   7.2 cm   3 cm   6.5 cm   10 cm

Q3   Work out the surface area of each shape:

a)   3 cm   5 cm   2 cm

b)   1 cm   3 cm   1 cm   2 cm   4 cm

c)   2 cm   2 cm   2 cm   4 cm   6 cm   10 cm

Q4   Work out the surface area of the cylinder.
Give your answer to 1 dp.

3 cm   5 cm

Stop nailing Daddy, Evil Kid.

Volume

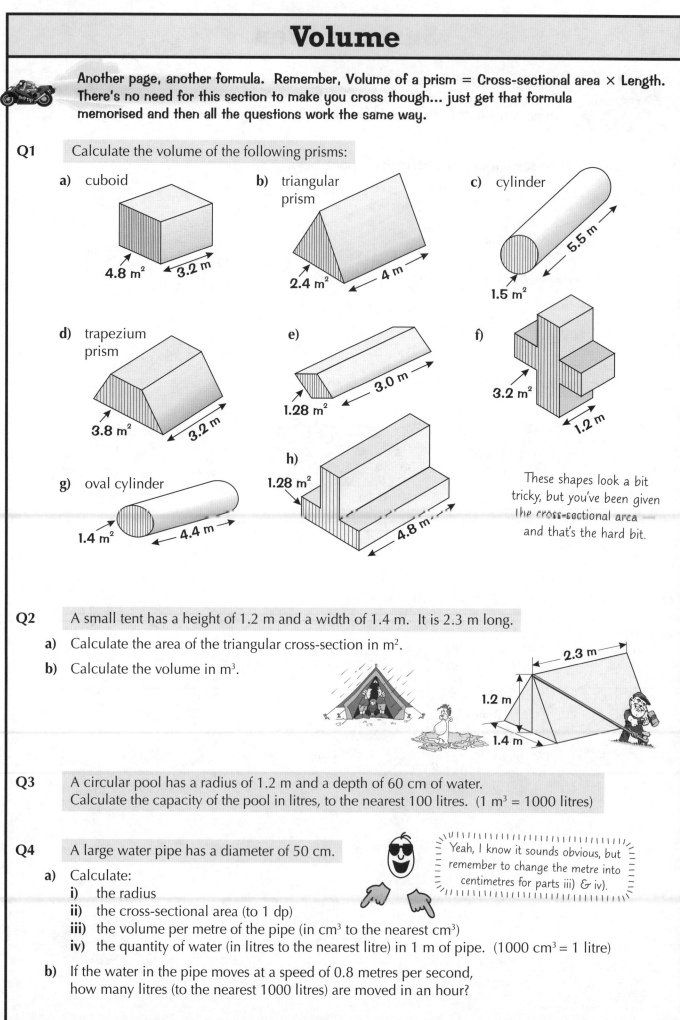

Another page, another formula. Remember, Volume of a prism = Cross-sectional area × Length. There's no need for this section to make you cross though... just get that formula memorised and then all the questions work the same way.

Q1 Calculate the volume of the following prisms:

a) cuboid

4.8 m²   3.2 m

b) triangular prism

2.4 m²   4 m

c) cylinder

5.5 m   1.5 m²

d) trapezium prism

3.8 m²   3.2 m

e)

3.0 m   1.28 m²

f)

3.2 m²   1.2 m

g) oval cylinder

1.4 m²   4.4 m

h)

1.28 m²   4.8 m

These shapes look a bit tricky, but you've been given the cross-sectional area — and that's the hard bit.

Q2 A small tent has a height of 1.2 m and a width of 1.4 m. It is 2.3 m long.

a) Calculate the area of the triangular cross-section in m².

b) Calculate the volume in m³.

2.3 m   1.2 m   1.4 m

Q3 A circular pool has a radius of 1.2 m and a depth of 60 cm of water. Calculate the capacity of the pool in litres, to the nearest 100 litres. (1 m³ = 1000 litres)

Q4 A large water pipe has a diameter of 50 cm.

Yeah, I know it sounds obvious, but remember to change the metre into centimetres for parts iii) & iv).

a) Calculate:
  i) the radius
  ii) the cross-sectional area (to 1 dp)
  iii) the volume per metre of the pipe (in cm³ to the nearest cm³)
  iv) the quantity of water (in litres to the nearest litre) in 1 m of pipe. (1000 cm³ = 1 litre)

b) If the water in the pipe moves at a speed of 0.8 metres per second, how many litres (to the nearest 1000 litres) are moved in an hour?

Angle Basics

Well here's acute page for you... There are loads of angle questions coming up, so it's good to get a little practice on the basics beforehand.

Q1 Describe each angle using the words 'acute', 'obtuse' or 'reflex'.

a)

b)

c)

d)

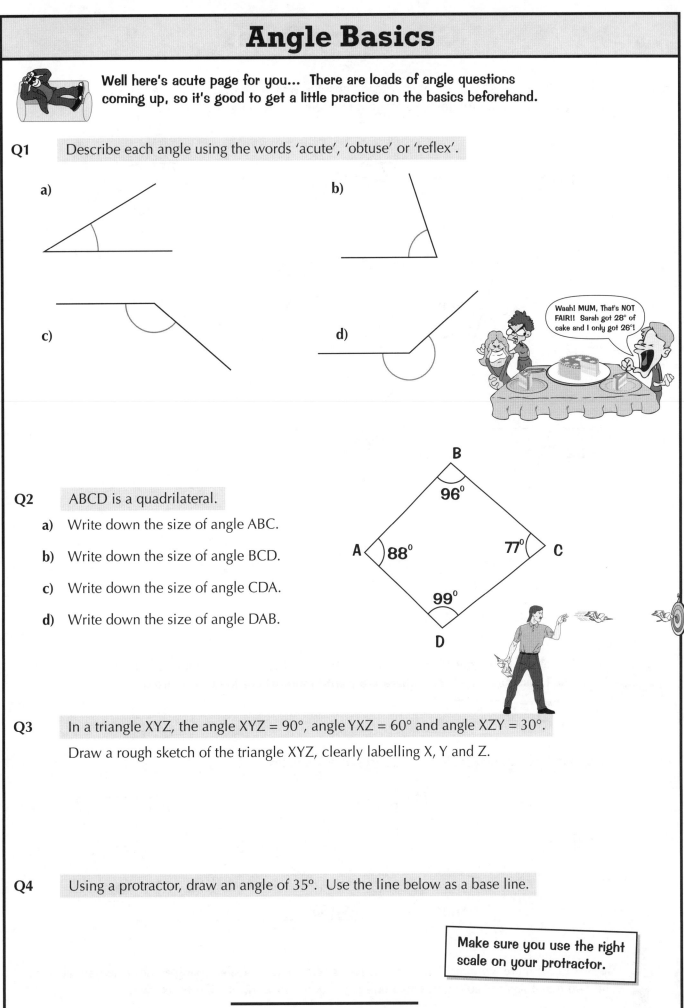

Waah! MUM, That's NOT FAIR!! Sarah got 28° of cake and I only got 26°!

Q2 ABCD is a quadrilateral.

a) Write down the size of angle ABC.

b) Write down the size of angle BCD.

c) Write down the size of angle CDA.

d) Write down the size of angle DAB.

B

96°

A 88° 77° C

99°

D

Q3 In a triangle XYZ, the angle XYZ = 90°, angle YXZ = 60° and angle XZY = 30°.

Draw a rough sketch of the triangle XYZ, clearly labelling X, Y and Z.

Q4 Using a protractor, draw an angle of 35°. Use the line below as a base line.

Make sure you use the right scale on your protractor.

Geometry Rules

Q1 Calculate the marked angles. Next to each answer, write one of the three reasons:
Angles round a point / Angles on a straight line / Vertically opposite angles.

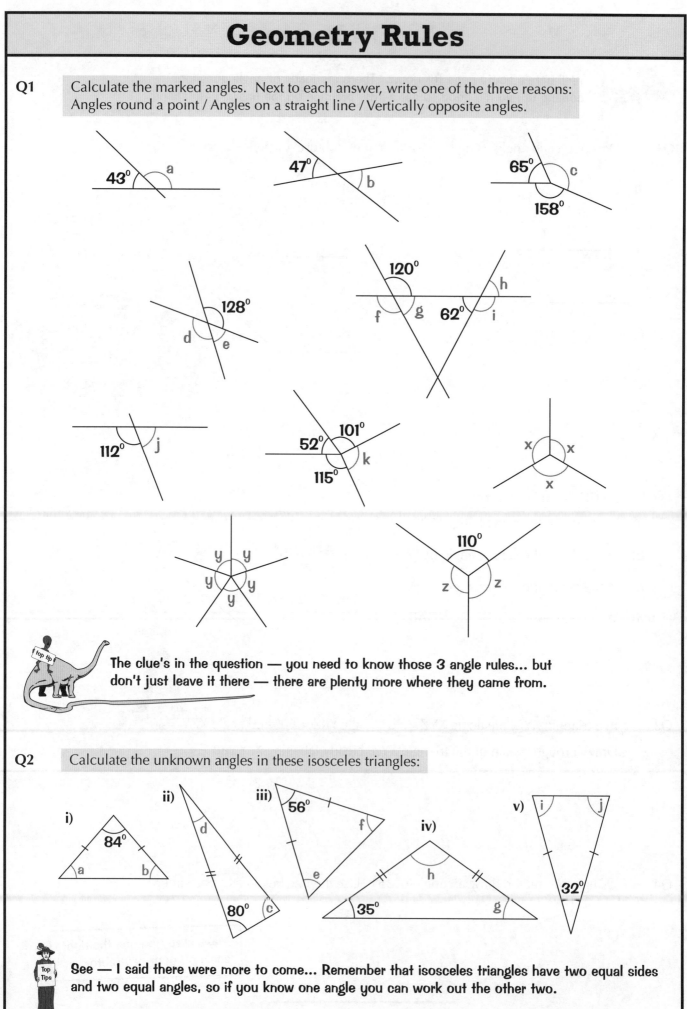

The clue's in the question — you need to know those 3 angle rules... but don't just leave it there — there are plenty more where they came from.

Q2 Calculate the unknown angles in these isosceles triangles:

See — I said there were more to come... Remember that isosceles triangles have two equal sides and two equal angles, so if you know one angle you can work out the other two.

Section 5 — Geometry and Measures

Geometry Rules

Q3   Work out the marked angles:

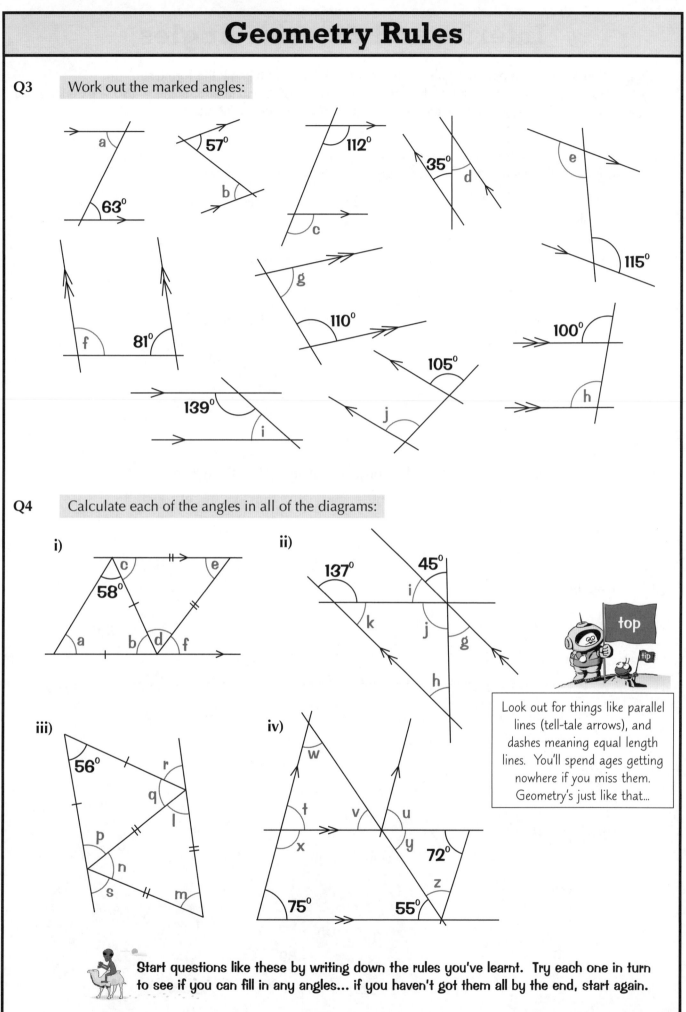

Q4   Calculate each of the angles in all of the diagrams:

i)

ii)

Look out for things like parallel lines (tell-tale arrows), and dashes meaning equal length lines. You'll spend ages getting nowhere if you miss them. Geometry's just like that...

iii)

iv)

Start questions like these by writing down the rules you've learnt. Try each one in turn to see if you can fill in any angles... if you haven't got them all by the end, start again.

Interior and Exterior Angles

You're bound to get asked about these, so make sure you know the formulas...
For any polygon, the sum of the exterior angles is 360°, and interior angle = 180° − exterior angle.
In a regular polygon the exterior angle is just 360° ÷ number of sides.

Q1 These diagrams show parts of regular polygons.
Calculate the interior and exterior angles in each case.

a) Regular decagon
(10 sides)

b) Regular 15-sided polygon

c) Regular 20-sided polygon

d) Regular 24-sided polygon

Q2 Find the number of sides of a regular polygon with an INTERIOR angle of:

    a) 140°   **b)** 150°   **c)** 160°   **d)** 168°   **e)** 170°

Q3 Find the missing angles:

For any polygon with n sides,
sum of interior angles = (n − 2) × 180°

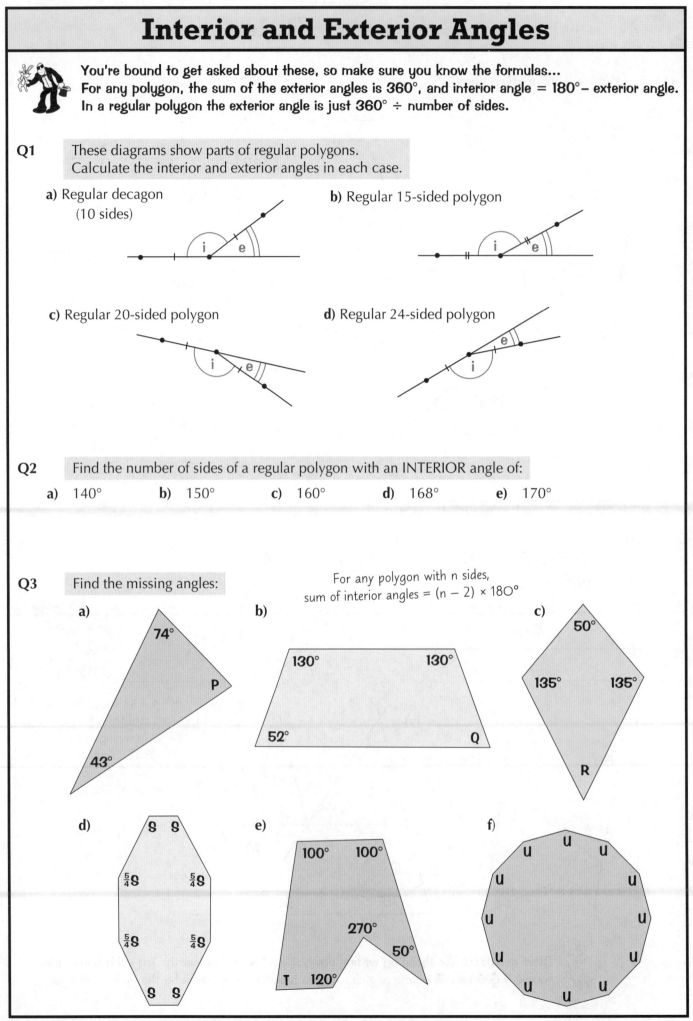

a) 74°, P, 43°

b) 130°, 130°, 52°, Q

c) 50°, 135°, 135°, R

d) s, s, $\frac{5}{4}$s, $\frac{5}{4}$s, $\frac{5}{4}$s, $\frac{5}{4}$s, s, s

e) 100°, 100°, 270°, 50°, T, 120°

f) u, u, u, u, u, u, u, u, u, u

Interior and Exterior Angles

Q4 The pentagon MNOPQ is shown.

a) Given that the exterior angle ONR is 110°
write down the size of angle MNO.

b) Angles NOP, OPQ and PQM are all the
same size as angle ONR. Work out the
size of angle QMN.

> Mark the angles you've worked out on the
> diagram as you go — it'll make life easier.

Q5 Find the missing angles A and B.

Q6 The following diagrams are not to scale. Calculate the marked angles.
In regular polygons, make use of the fact that all the triangles are isosceles.

i)

ii)

> The angle at the centre of a
> regular polygon is the same
> size as the exterior angle.

Regular Pentagon

Regular Hexagon

iii)

iv)

**Pentagon with one
line of symmetry**

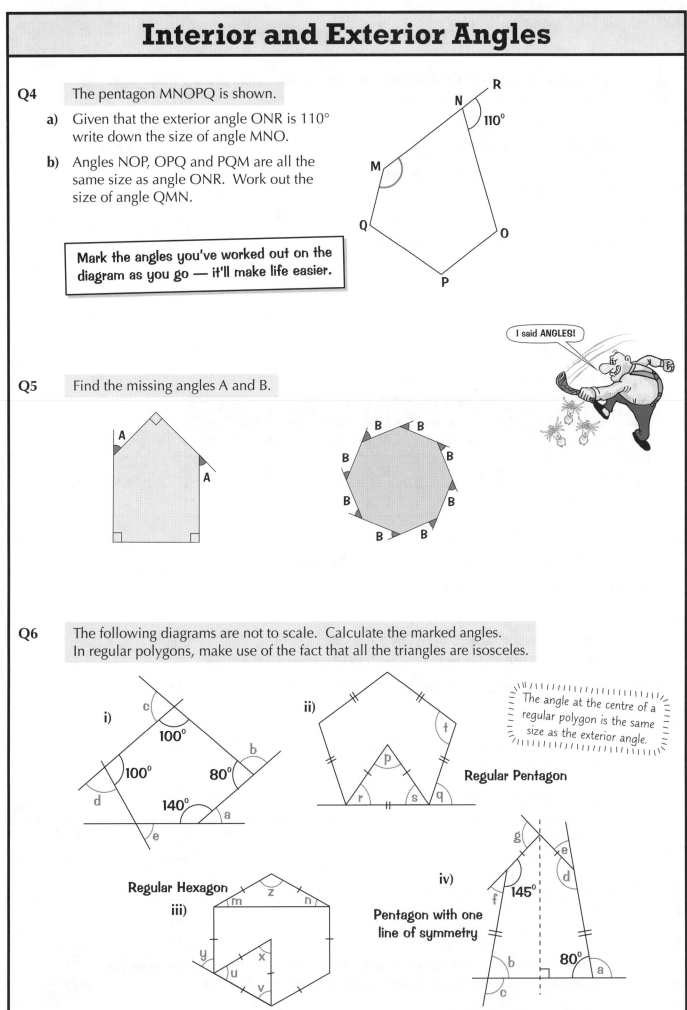

Transformations

It's a good idea to learn the 4 different transformations before starting these pages — they are translation, reflection, rotation and enlargement.

Q1 On graph paper draw axes from -6 to +6 in each direction.

a) Copy the shape OABC and its reflection in the *y*-axis, OA′B′C′.

b) Draw the reflection of OABC in the *x*-axis and label it OA″B″C″.

c) Finally draw the reflection of OA′B′C′ in the *x*-axis and label it OA‴B‴C‴.

d) Copy and complete the chart of coordinates:

Original	Reflection (a)	Reflection (b)	Reflection (c)
A (6, 2)	A′ (-6, 2)	A″ (6, -2)	A‴ (-6, -2)
B ( )	B′ ( )	B″ ( )	B‴ ( )
C ( )	C′ ( )	C″ ( )	C‴ ( )

e) Copy and complete these statements:

i) When reflecting in the *x*-axis, the sign of the *y* coordinate changes but ....

ii) When reflecting in the *y*-axis, ... but ....

Q2 B is the translation of the triangle A by the vector $\begin{pmatrix} 9 \\ 2 \end{pmatrix}$
(9 units in the positive *x*-direction, 2 units in the positive *y*-direction).

a) Copy the diagram onto graph paper.

b) Translate A by the vector $\begin{pmatrix} 6 \\ -4 \end{pmatrix}$. Label the result C.

c) Translate A now by the vector $\begin{pmatrix} 0 \\ -6 \end{pmatrix}$
(This will be a movement parallel to the *y*-axis).
Label the result D.

Make sure you get the vector numbers the right way round...

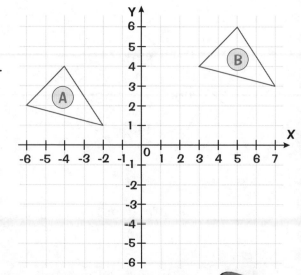

**Yikes — vectors... well no-one likes these, but you're still gonna have to do them.
They represent a certain distance in a certain direction, that's all.**

Transformations

Q3   In this question the transformations are either reflections or rotations about (0, 0). Describe in full the transformations which take:

a) P to Q

b) Q to R

c) R to S

d) S to P

e) R to P

f) Q to S

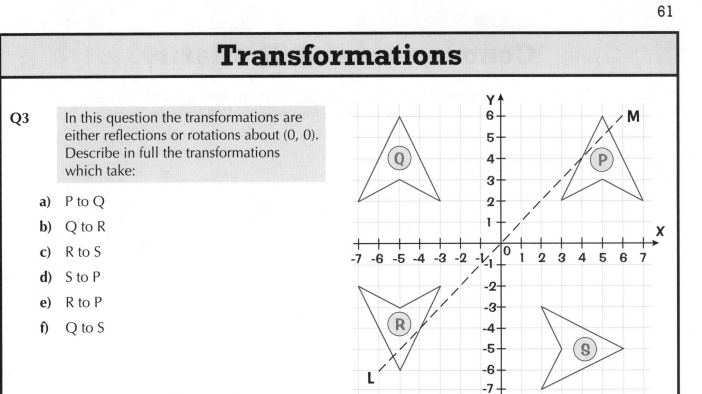

 The best way to figure out a rotation is to get out your trusty tracing paper.

Q4   Copy and complete the chart, using the origin as the centre of enlargement, then complete the diagram.

Original	Enlargement×2	Enlargement×3
A (3, 0)	A' ( )	A" ( )
B (3, 1)	B' ( )	B" ( )
C (1½, 2)	C' ( )	C" ( )
D (0, 1)	D' ( )	D" ( )

Remember — a scale factor of 2 means each point on the new shape is twice as far from the centre of enlargement as the same point on the original shape.

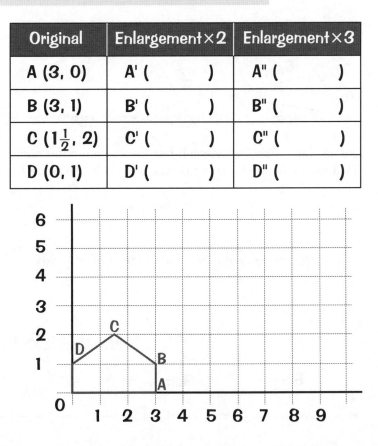

Congruence and Similarity

If you're not sure whether two shapes are <u>exactly</u> the same, tracing paper's always a good idea — then you can put one on top of the other and it'll be easier to see.

Q1 Find the pairs of congruent shapes.

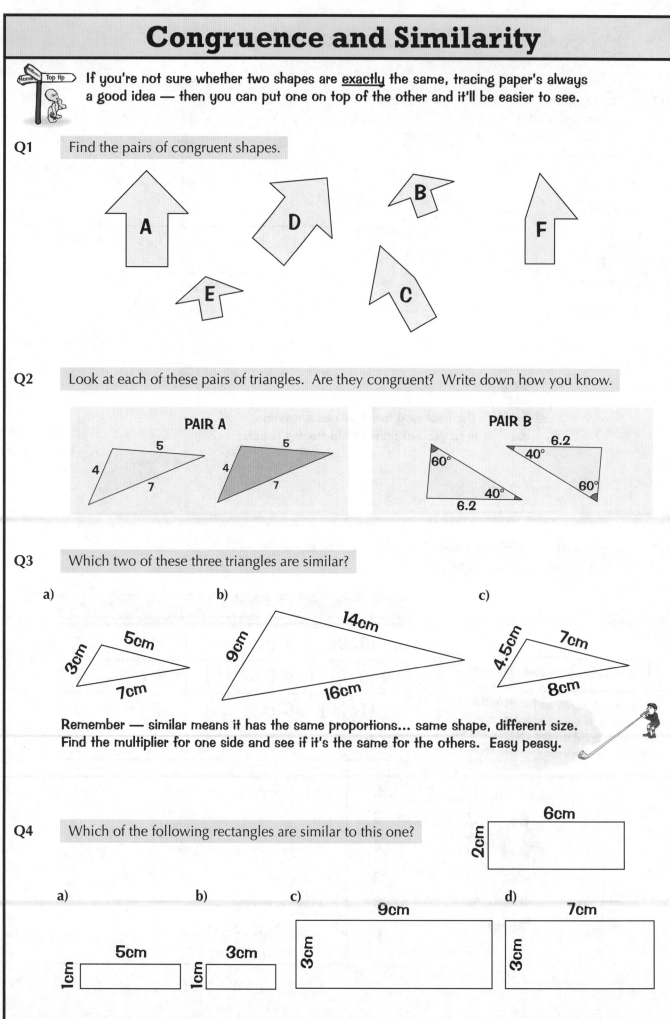

Q2 Look at each of these pairs of triangles. Are they congruent? Write down how you know.

PAIR A

PAIR B

Q3 Which two of these three triangles are similar?

a)

b)

c)

Remember — similar means it has the same proportions... same shape, different size. Find the multiplier for one side and see if it's the same for the others. Easy peasy.

Q4 Which of the following rectangles are similar to this one?

6cm
2cm

a)
5cm
1cm

b)
3cm
1cm

c)
9cm
3cm

d)
7cm
3cm

Section 5 — Geometry and Measures

Constructions

Umm, no, put away your hammer and chisel. All you need for these constructions is a protractor, a ruler, your compasses and a nice sharp pencil.

Q1   Using compasses and a ruler construct these triangles.

a)   triangle ABC with AB = 7 cm, BC = 5 cm, CA = 3 cm.

b)   triangle DEF with DE = 8 cm, EF = 5.5 cm, FD = 6 cm.

c)   triangle GHI with GH = HI = IG = 6.5 cm.

d)   triangle JKL with JK = 9.3 cm, KL = 4.5 cm, LJ = 7.5 cm.

Q2   Draw accurately each of these triangles.

a)   triangle MNP with angle M = 30°, MN = 6 cm, MP = 5 cm.

b)   triangle QRS with angle R = 70°, QR = 9.3 cm, SR = 5.5 cm.

c)   triangle TUV with angle V = 50°, VT = 7 cm, UV = 9 cm.

d)   triangle WXY with angle X = 80°, XW = XY = 7 cm.

   What do you notice about triangles TUV and WXY?

Q3   Draw accurately each of these triangles.

a)   triangle PTO with angle P = 30°, angle T = 60°, PT = 7 cm.

b)   triangle TLC with angle T = 40°, angle C = 35°, TC = 6.6 cm.

c)   triangle GSH with angle S = 112°, angle H = 35°, SH = 8.3 cm.

d)   triangle PDQ with angle P = 28°, angle D = 39°, PD = 5.8 cm.

Q4   Construct triangle ABC accurately with length AB = 10 cm, angle ABC = 48° and angle CAB = 25°.

You'll need your compasses for this question.

a)   Construct the perpendicular bisector of the line AC. Mark point K where the bisector crosses the line AB.

b)   Bisect angle ACB. Mark point J where the bisector crosses the line AB. Measure the length JK.

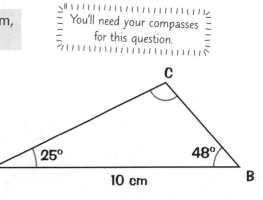

Pythagoras' Theorem

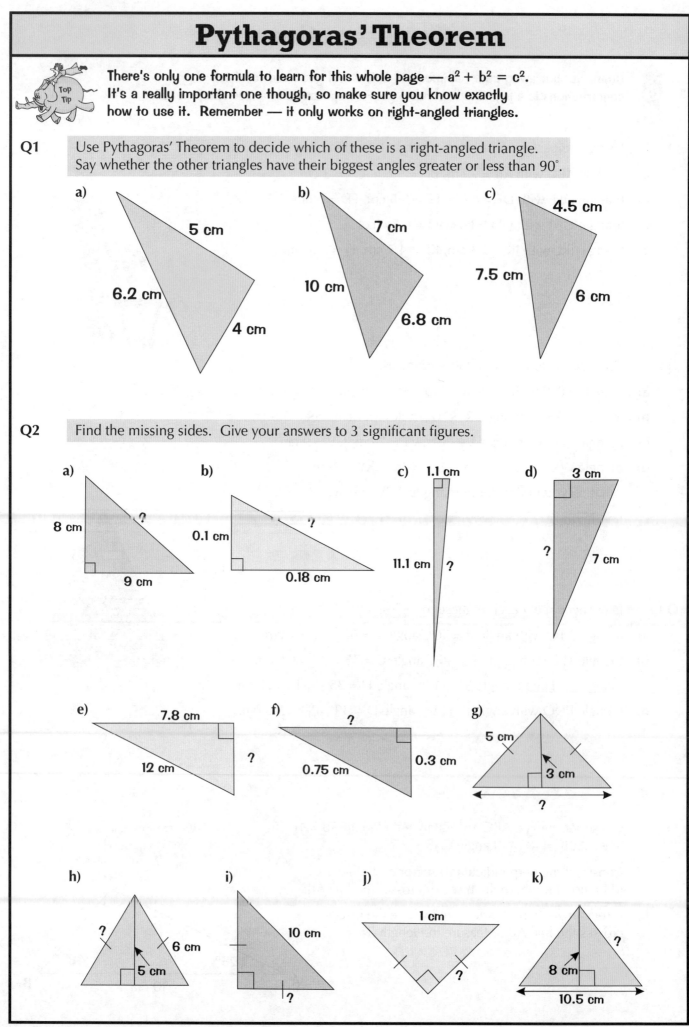

There's only one formula to learn for this whole page — $a^2 + b^2 = c^2$.
It's a really important one though, so make sure you know exactly
how to use it. Remember — it only works on right-angled triangles.

Q1 Use Pythagoras' Theorem to decide which of these is a right-angled triangle.
Say whether the other triangles have their biggest angles greater or less than 90°.

a) 5 cm   6.2 cm   4 cm

b) 7 cm   10 cm   6.8 cm

c) 4.5 cm   7.5 cm   6 cm

Q2 Find the missing sides. Give your answers to 3 significant figures.

a) 8 cm   ?   9 cm

b) 0.1 cm   ?   0.18 cm

c) 1.1 cm   11.1 cm   ?

d) 3 cm   ?   7 cm

e) 7.8 cm   12 cm   ?

f) ?   0.75 cm   0.3 cm

g) 5 cm   3 cm   ?

h) ?   6 cm   5 cm

i) 10 cm   ?

j) 1 cm   ?

k) ?   8 cm   10.5 cm

Trigonometry

Here it is — the big daddy of the geometry world. It might be difficult to get your head round at first, but it's worth persevering. You may as well write down the 'SOH', 'CAH', 'TOA' triangles now — they'll come in dead handy over the next couple of pages.

Q1 Copy and complete this table using your calculator. Give your answers to 3 decimal places.

Angle a	Cos a	Sin a	Tan a
0°	1		
10°			
15°			
		0.5	
45°			
	0.5		
80°			
88°			
90°			

Q2 Label each side as "hypotenuse", "adjacent" and "opposite" relative to the angle marked:

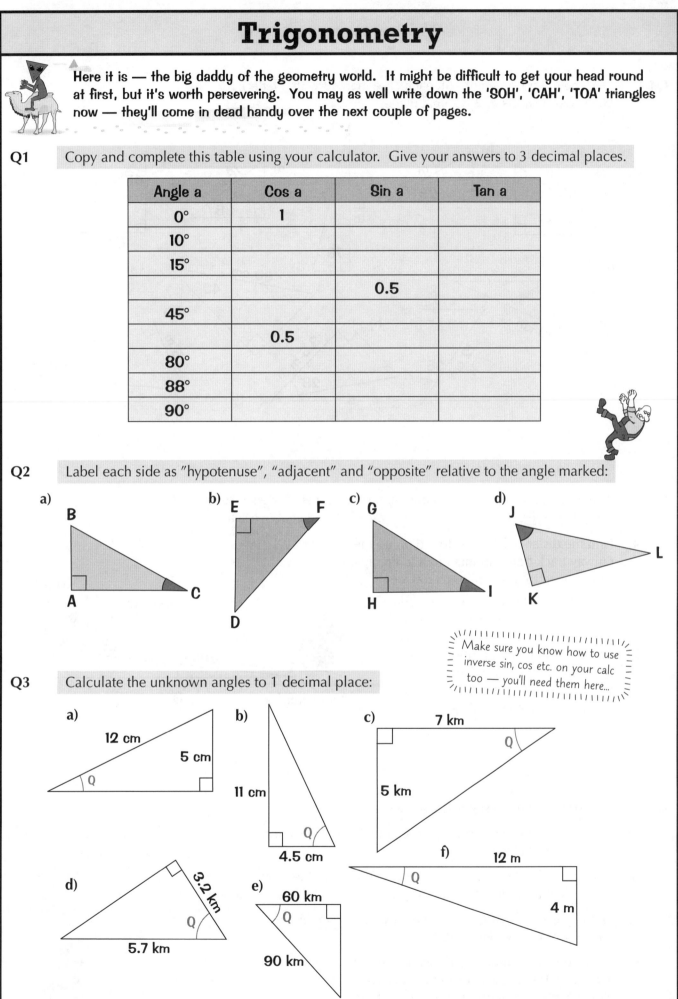

Make sure you know how to use inverse sin, cos etc. on your calc too — you'll need them here...

Q3 Calculate the unknown angles to 1 decimal place:

Trigonometry

Q4   Find the unknown lengths (in each case marked with a letter):
Give your answers to 2 decimal places.

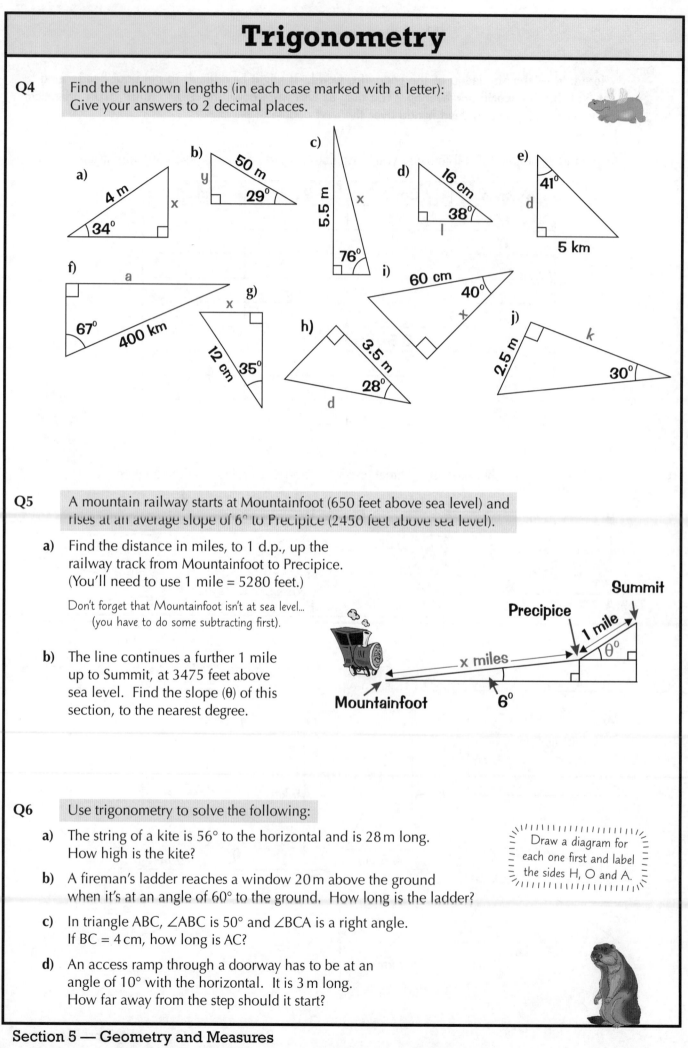

Q5   A mountain railway starts at Mountainfoot (650 feet above sea level) and
rises at an average slope of 6° to Precipice (2450 feet above sea level).

a)   Find the distance in miles, to 1 d.p., up the
railway track from Mountainfoot to Precipice.
(You'll need to use 1 mile = 5280 feet.)

Don't forget that Mountainfoot isn't at sea level...
(you have to do some subtracting first).

b)   The line continues a further 1 mile
up to Summit, at 3475 feet above
sea level. Find the slope (θ) of this
section, to the nearest degree.

Q6   Use trigonometry to solve the following:

a)   The string of a kite is 56° to the horizontal and is 28 m long.
How high is the kite?

b)   A fireman's ladder reaches a window 20 m above the ground
when it's at an angle of 60° to the ground. How long is the ladder?

c)   In triangle ABC, ∠ABC is 50° and ∠BCA is a right angle.
If BC = 4 cm, how long is AC?

d)   An access ramp through a doorway has to be at an
angle of 10° with the horizontal. It is 3 m long.
How far away from the step should it start?

Draw a diagram for
each one first and label
the sides H, O and A.

Probability

Probability is one of my favourite topics — you find out all kinds of interesting things. To find the probability of something, divide the number of ways it can happen by the total number of possible outcomes. Try these questions and you'll get the idea.

Q1 An ordinary pack of 52 cards is shuffled and you choose a card at random. Give the probability that the card is:

a) red

b) a heart

c) a 3

d) a court card (Jack, Queen or King)

e) not a court card

f) an even number

g) a number less than 6
(assuming an Ace is higher than a 6)

h) an Ace

i) the Ace of Spades.

Give your answers as fractions in their simplest forms.

Probabilities are always between 0 and 1, so they're sometimes given as fractions and sometimes as decimals.

Q2 Write down, as a fraction, the probability of these events happening:

a) throwing a 5 with a dice

b) throwing an even number with a dice

c) throwing a prime number with a dice

d) throwing a 0 with a dice.

The evil dice hoards will never take CleaverGirl alive.

Q3 A bag contains ten balls. Five are red, three are yellow and two are eyeballs. What is the probability of picking out:

a) a yellow ball

b) a red ball

c) an eyeball

d) a red ball or an eyeball

e) a blue ball

Q4 In a party game, tickets numbered from 1 to 100 are put in a hat and have to be picked out blindfolded. You win a prize if you pick a ticket number ending in 7.

a) What is P(win), assuming you can only pick once?

b) What is P(not win), assuming you can only pick once?
Give your answers as fractions in their simplest forms.

Probability

Q5   Spinner A is divided into 4 equal parts: red, black, white and pink.
Spinner B is divided into 3 equal parts: white, red and black.

a)   Complete the chart below showing all possible outcomes:

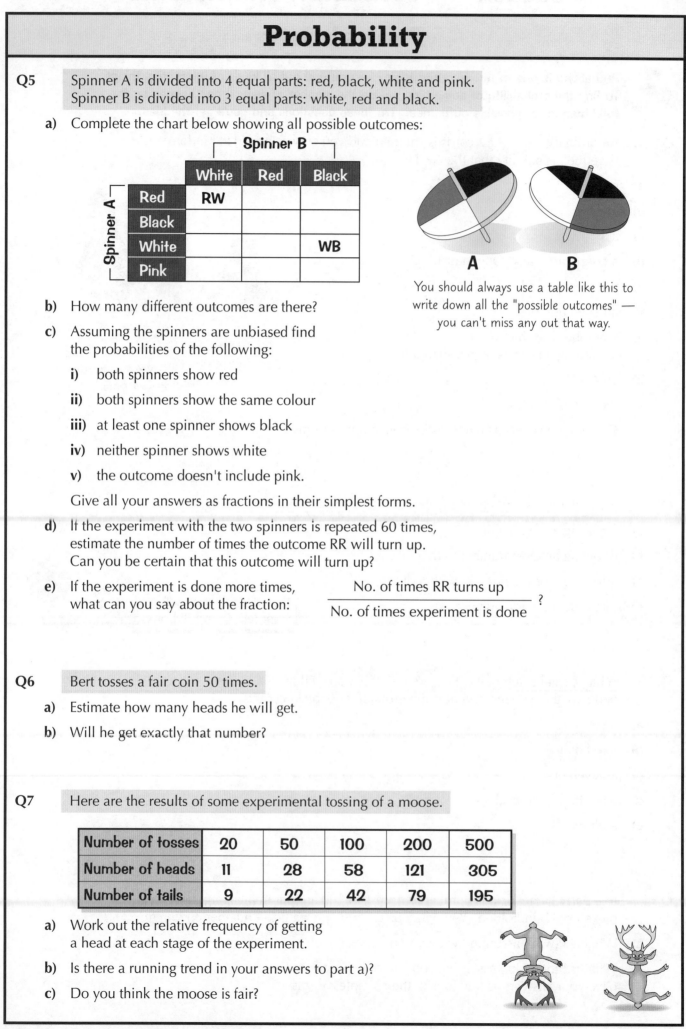

Spinner B

		White	Red	Black
Spinner A	Red	RW		
	Black			
	White			WB
	Pink			

You should always use a table like this to write down all the "possible outcomes" — you can't miss any out that way.

b)   How many different outcomes are there?

c)   Assuming the spinners are unbiased find the probabilities of the following:

    i)   both spinners show red

    ii)   both spinners show the same colour

    iii)   at least one spinner shows black

    iv)   neither spinner shows white

    v)   the outcome doesn't include pink.

Give all your answers as fractions in their simplest forms.

d)   If the experiment with the two spinners is repeated 60 times, estimate the number of times the outcome RR will turn up. Can you be certain that this outcome will turn up?

e)   If the experiment is done more times, what can you say about the fraction:

$$\frac{\text{No. of times RR turns up}}{\text{No. of times experiment is done}} ?$$

Q6   Bert tosses a fair coin 50 times.

a)   Estimate how many heads he will get.

b)   Will he get exactly that number?

Q7   Here are the results of some experimental tossing of a moose.

Number of tosses	20	50	100	200	500
Number of heads	11	28	58	121	305
Number of tails	9	22	42	79	195

a)   Work out the relative frequency of getting a head at each stage of the experiment.

b)   Is there a running trend in your answers to part a)?

c)   Do you think the moose is fair?

Section 6 — Probability and Statistics

Venn Diagrams

Don't be put off by the weird symbols that go with the sets — once you've figured them out you'll find that Venn diagrams are a pretty handy tool. Remember, the squiggly symbol ξ means the 'universal set' — that contains all the elements that you need to consider in a particular question.

Q1 The elements of ξ are the odd numbers between 0 and 16.
Set P contains the multiples of 3, and set Q contains the multiples of 5.

a) Using set notation, write sets ξ, P and Q as complete lists of their elements.

b) Show the elements of these sets on a copy of the Venn diagram below.

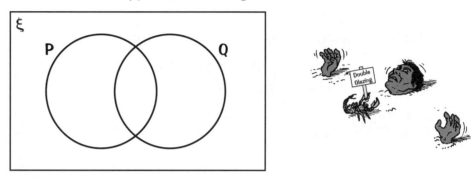

Q2 Using the Venn diagram on the right, write down the following:

a) The elements of set S.

b) The elements of set R $\cap$ S.

c) n(R)

d) The elements of set S'.

e) n(R')

f) n(R $\cup$ S)

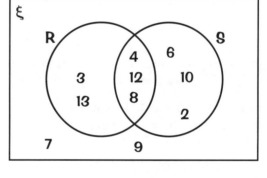

Q3 A survey asked people if they like cats and dogs.
8 people said they only like cats, 9 people said they only like dogs,
4 people said they like both and 6 people said they like neither.

a) Draw a Venn diagram to show this data.

b) A person from the survey is chosen at random.
What is the probability that the chosen person:

i) likes cats

ii) doesn't like dogs?

Q4 ξ = {integers from 1-16 inclusive}, A = {multiples of 2} and B = {square numbers}.

a) What are the elements of A $\cap$ B?

b) Draw a Venn diagram, showing the number of elements in each part of the diagram.

c) If a card from a set of cards numbered 1-16 is chosen at random, what is the probability that:

i) the card is a multiple of 2

ii) the card is both a multiple of 2 and a square number

iii) the number is not a square number?

Graphs and Charts

Pie, bar... all this talk of grub is making me hungry. Make sure you know the golden rule of pie charts for this page — the total of everything = 360°.

Q1 Ten families live in Leafy Close. 1 family has no children, 2 families have 1 child each, 3 families have 2 children, 1 has 3 children, 2 have 4 children and 1 has 5 children.

a) Are we dealing with discrete or continuous data?

b) Plot the information on a bar chart.

Q2 Here is a bar chart showing pet ownership in year 7 of 2 different schools:

a) Which school's pupils have more pets in total?

b) Which school has more cats?

c) How many more rabbits do the pupils of school A have than those of school B?

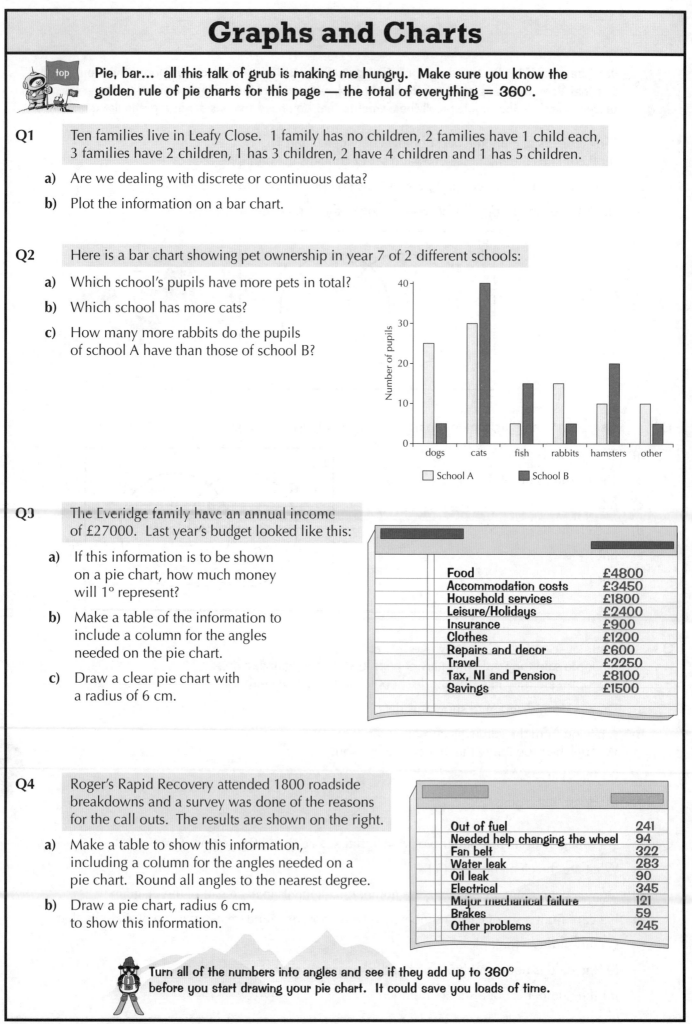

Q3 The Everidge family have an annual income of £27000. Last year's budget looked like this:

a) If this information is to be shown on a pie chart, how much money will 1° represent?

b) Make a table of the information to include a column for the angles needed on the pie chart.

c) Draw a clear pie chart with a radius of 6 cm.

Food	£4800
Accommodation costs	£3450
Household services	£1800
Leisure/Holidays	£2400
Insurance	£900
Clothes	£1200
Repairs and decor	£600
Travel	£2250
Tax, NI and Pension	£8100
Savings	£1500

Q4 Roger's Rapid Recovery attended 1800 roadside breakdowns and a survey was done of the reasons for the call outs. The results are shown on the right.

a) Make a table to show this information, including a column for the angles needed on a pie chart. Round all angles to the nearest degree.

b) Draw a pie chart, radius 6 cm, to show this information.

Out of fuel	241
Needed help changing the wheel	94
Fan belt	322
Water leak	283
Oil leak	90
Electrical	345
Major mechanical failure	121
Brakes	59
Other problems	245

Turn all of the numbers into angles and see if they add up to 360° before you start drawing your pie chart. It could save you loads of time.

Graphs and Charts

Q5 Here is a frequency bar chart showing the heights of 60 Brussels sprout plants in Eric's garden.

a) How many plants were less than 15 cm high?

b) How many plants were more than 25 cm high?

c) Can you tell how many plants were between 20 and 23 cm high?

Heights of Eric's Brussels Sprouts

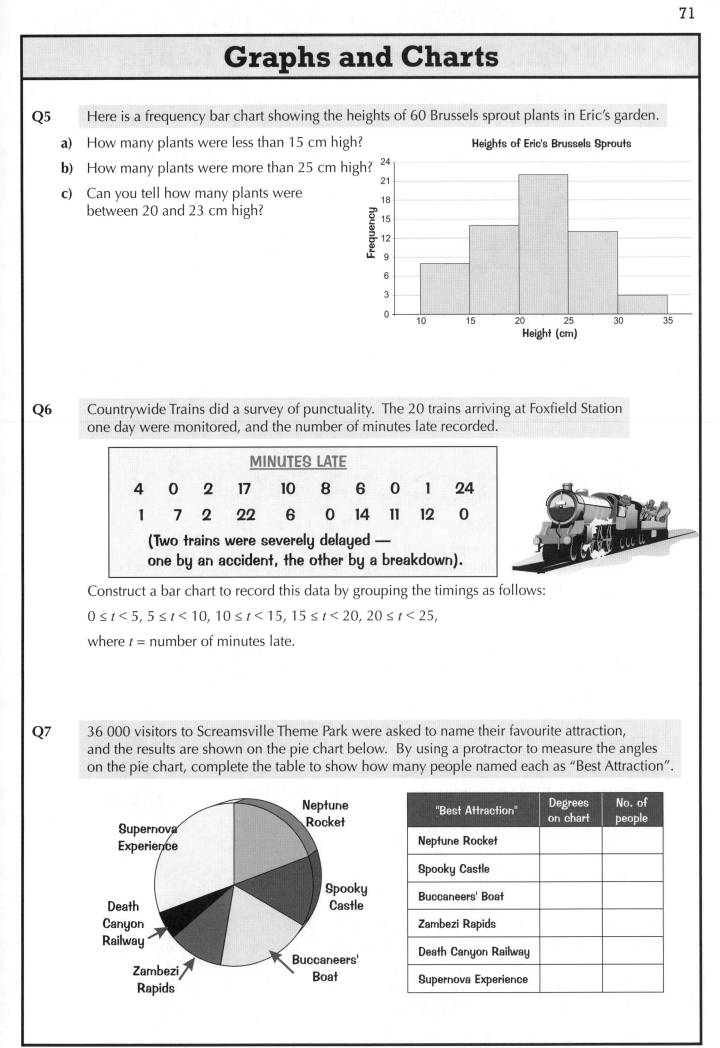

Q6 Countrywide Trains did a survey of punctuality. The 20 trains arriving at Foxfield Station one day were monitored, and the number of minutes late recorded.

MINUTES LATE

| 4 | 0 | 2 | 17 | 10 | 8 | 6 | 0 | 1 | 24 |
| 1 | 7 | 2 | 22 | 6 | 0 | 14 | 11 | 12 | 0 |

**(Two trains were severely delayed —
one by an accident, the other by a breakdown).**

Construct a bar chart to record this data by grouping the timings as follows:

$0 \leq t < 5$, $5 \leq t < 10$, $10 \leq t < 15$, $15 \leq t < 20$, $20 \leq t < 25$,

where t = number of minutes late.

Q7 36 000 visitors to Screamsville Theme Park were asked to name their favourite attraction, and the results are shown on the pie chart below. By using a protractor to measure the angles on the pie chart, complete the table to show how many people named each as "Best Attraction".

"Best Attraction"	Degrees on chart	No. of people
Neptune Rocket		
Spooky Castle		
Buccaneers' Boat		
Zambezi Rapids		
Death Canyon Railway		
Supernova Experience		

Mean, Median, Mode and Range

This page is full of all sorts of different averages questions. Unfortunately you'll have learn how each type of average works, because these are bound to come up in tests. I know, I'm mean...

Q1 Find the mode and median shoe size of 30 school children whose shoe sizes are:

4	2	3	4	4	5	4	2	4	3
2	1	3	1	3	2	5	3	2	3
3	2	3	2	4	6	7	2	3	1

Don't forget, put the data into ascending order before you start.

Q2 Find the median, mode, mean and range of the following data:

a) 15, 13, 11, 9, 7, 10, 0, 0, 1, 3

b) 6, 2, 3, 3, 5, 4, 4, 5, 3

c) 8, 8, 8, 8, 8

d) 1, 3, 2, 4, 2, 4, 2, 4, 2

Q3 On a large box of maggots it says "Average contents 240 maggots".
I counted the number of maggots in ten boxes. These are the results:

241   244   236   240   239   242   237   239   239   236

Is the label on the box correct?
Use the mean, median and mode for the numbers of maggots to explain your answer.

Q4 Gary has a Saturday job. He earns £4.20 an hour. He thinks that most of his friends earn more. Here is a list of how much his friends are paid per hour.

Lisa   £4.10   Scott   £5.00

Kate   £4.75   Kylie   £3.90

Helen   £4.51   Kirsty   £4.75

Ben   £4.75   Ruksana   £4.40

By working out the mean, median and mode for his friends' pay, investigate whether Gary is right.

Q5 The table below shows the number of absences for each pupil in a class during a school year.

a) Calculate the mean number of absences per pupil.

b) List the data in order and find the median number of absences per pupil.

c) How many pupils have a number of absences above the mean, and how many are below it?

ABSENCES				
0	1	0	0	5
18	3	2	2	3
0	0	1	1	4
3	2	5	20	0
1	2	24	2	3

Averages from Frequency Tables

 Sorted out your averages yet? Make sure you're confident with the basics of the mean, median and mode before you start on this section — things get a bit more awkward when frequency tables are involved.

Q1   Use this frequency table to work out the mean, median, mode and range for the number of spots on 30 different dogs.

Number of spots	10	11	12	13	14	15	16	17	18	19	20
Frequency	1	0	3	7	8	5	3	0	2	0	1

Q2   Use this frequency table to work out the mean, median, mode and range for the number of spikes on 25 different styles of leather dog-collar.

Number of spikes	8	9	10	11	12	13	14	15	16
Frequency	4	2	5	4	1	4	3	0	2

Q3   Uptown School gives their GCSE Maths candidates a revision book called "Never Mind the Maths", while Downshire School use a book called "The Sum has got its Hat On". Pupils in each school study their revision books for 3 months, then take mock exams. The results are summarised below:

UPTOWN SCHOOL

Score (out of 100)	1 - 20	21 - 40	41 - 60	61 - 80	81 - 100
No. of pupils	13	32	68	63	29
Mid-Interval					
No. of pupils × Mid-Interval					

DOWNSHIRE SCHOOL

Score (out of 100)	1 - 20	21 - 40	41 - 60	61 - 80	81 - 100
No. of pupils	15	48	76	40	18
Mid-Interval					
No. of pupils × Mid-Interval					

a)   Using a scale of 1 cm to 10 marks horizontally and 1 cm to 10 pupils vertically, draw two frequency polygons on the same axes, to represent the data.

Use different colours for each frequency polygon, so you don't get muddled up.

b)   Using mid-interval values (10.5, 30.5, 50.5, etc) estimate the mean score for each school.

c)   Does this prove that "Never Mind the Maths" is a better book than "The Sum has got its Hat On"? Explain.

Averages from Frequency Tables

Q4 The age profile of the population of Northland is contained in the frequency table below:

Age (whole number of years)	Frequency	Mid-Interval	Frequency × Mid-Interval
0 - 9	107	4.5	107 × 4.5 = 481.5
10 - 19	130	14.5	130 × 14.5 = 1885
20 - 29	132	24.5	
30 - 39	144		
40 - 49	120		
50 - 59	113		
60 - 69	110		
70 - 79	90		
80 - 89	49		
90 - 99	5		

a) Complete the table and use it to calculate an estimated mean age for the population (to 1 dp).

b) Which interval contains the median value?

c) Which is the modal class?

> The mid-interval value method for estimating the mean is definitely worth having under your belt — it's a classic test question.

Q5 Another country, Southland has a very different age profile:

Age (whole number of years)	Frequency	Mid-Interval	Frequency × Mid-Interval
0 - 9	272		
10 - 19	227		
20 - 29	173		
30 - 39	124		
40 - 49	99		
50 - 59	60		
60 - 69	29		
70 - 79	8		
80 - 89	6		
90 - 99	2		

a) Use the mid-interval technique to estimate the mean age of the population.

b) Which interval contains the median age of the population?

c) Which is the modal class?

Averages from Frequency Tables

Q6 A small company has 93 employees, and their wages are summarised in the table:

Wage (£)	No. of Employees	Mid-Interval	Frequency × Mid-Interval
0 ⩽ W < 5000	0		
5000 ⩽ W < 10000	29		
10000 ⩽ W < 15000	31		
15000 ⩽ W < 20000	16		
20000 ⩽ W < 25000	8		
25000 ⩽ W < 30000	5		
30000 ⩽ W < 35000	2		
35000 ⩽ W < 40000	2		

a) Complete the table and use the mid-interval technique to estimate the mean wage (to the nearest £100)

b) In which class interval does the median value lie?

c) Which is the modal class?

Here we go again — more practice using the mid-interval value method... you've really got to be able to do these, so keep trying until you can.

Q7 A different company has 100 employees and a wage profile as follows:

Wage (£)	No. of Employees	Mid-Interval	Frequency × Mid-Interval
0 ⩽ W < 5000	0		
5000 ⩽ W < 10000	2		
10000 ⩽ W < 15000	30		
15000 ⩽ W < 20000	28		
20000 ⩽ W < 25000	17		
25000 ⩽ W < 30000	10		
30000 ⩽ W < 35000	5		
35000 ⩽ W < 40000	4		
40000 ⩽ W < 45000	2		
45000 ⩽ W < 50000	2		

a) Complete the table showing mid-interval values for the wages, and use it to estimate the mean wage.

b) In which interval does the median wage lie?

c) Which is the modal class?

Scatter Graphs

Here it is, the last page of questions. What do you mean "yay"? Well, keep going to the end. Scatter graphs are good for showing you whether two things are linked — the posh word for this is 'correlated'. If they are, you can draw a line of best fit through your points.

Q1 Draw a scatter graph using the following data. It shows the number of customers and the takings in a clothes shop on eight different days.

Number of customers	200	120	320	400	60	330	280	250
Takings (£1000)	4.2	2	4.5	8	1.1	6	4.8	5.1

My best mate's a badger,
He wears a big moustache,
He spits real far
At the Opera
And he belches with panache...

a) Draw the line of best fit through your points.

b) What kind of correlation does your graph show?

c) Use your line to predict the takings on a day when there were 260 customers.

d) Use your line to predict the number of customers on a day when the takings were £6500.

Q2 This data shows the wages earned and hours worked by nine year 12 students who have Saturday jobs.

	A	B	C	D	E	F	G	H	I
Hours	2	1.5	3	6	5	4.5	2.5	4.25	3.5
Wages (£)	8	5	9	21	17.50	17	10	15	12

a) Draw a scatter graph for this data.

b) Draw the line of best fit.

c) Use your line to predict how much a student working 5 hours might earn.

d) How many hours might a student have to work to earn £16?

Heyyy....*

Q3 This graph shows the size of a car's engine and the distance it can travel on 1 litre of fuel, for several different models of car.

a) Draw a line of best fit.

b) Describe the relationship between engine size and the number of miles a car can travel per litre. What kind of correlation is this?

c) Use your graph to predict the distance a car with an engine capacity of 1.8 litres can travel per litre of fuel.

*We would like to apologise for the poor quality of the graphics on this page. The Quality Control Department is pleased to announce that the fool responsible has since been severely reprimanded. And given a wedgie.

Answers

Section 1 — Numbers

Page 1 — Calculating Tips

Q1 a) no (ans = 6)
   b) "12 ÷ 4 × 0.5 ="
   c) and d) "12 ÷ (4 × 0.5)"
      OR "4 × 0.5 ="
      then "12 ÷ Ans ="
      OR "12 ÷ 4 ÷ 0.5"

Q2 "140 ÷ (7 + 13) ="
   OR "7 + 13 ="
   then "140 ÷ Ans ="

Q3 a) 3.8   **e)** 150
   b) 5.5   **f)** -1
   c) 18.9   **g)** 9.9
   d) 158.6   **h)** 0.387 (3 dp)

Q4 Possible answers: -16, -4, 8, 13, 16, 19, 40.

Page 2 — Ordering Numbers

Q1 a) 9741 (nine thousand seven hundred and forty-one). 1479 (one thousand four hundred and seventy-nine).
   b) 8843 (eight thousand eight hundred and forty-three) 3488 (three thousand four hundred and eighty-eight)
   c) 9432 (nine thousand four hundred and thirty-two). 2349 (two thousand three hundred and forty-nine).
   d) 85443 (eighty-five thousand four hundred and forty-three) 34458 (thirty-four thousand four hundred and fifty-eight)
   e) 87321 (eighty-seven thousand three hundred and twenty-one). 12378 (twelve thousand three hundred and seventy-eight).
   f) 97321 (ninety-seven thousand three hundred and twenty-one). 12379 (twelve thousand three hundred and seventy-nine).

Q2 a) 8 units (8)
   b) 8 tens or eighty (80)
   c) 8 tenths (0.8)
   d) 8 hundredths (0.08)
   e) 80 thousands or 8 ten-thousands (80 000)
   f) 8 thousandths (0.008)
   g) 8 hundreds (800)
   h) 8 hundred thousands (800 000)
   i) 8 thousands (8 000)

Q3 a) 1.3, 1.5, 1.54, 1.62, 1.71, 1.89, 1.98
   b) 100.3, 100.4, 100.43, 101.2, 101.6, 102.8, 102.89
   c) -10, -3, -1, 0, 2, 4, 5
   d) 7.09, 7.13, 7.18, 7.21, 7.36, 7.40, 7.41

Q4 a) 4.1 cm, 4.0 cm, 3.9 cm, 3.1 cm, 2.3 cm, 2 cm, 0.9 cm
   b) 79.1 km, 78.7 km, 76.1 km, 75.2 km, 74.9 km, 74.3 km, 74.1 km
   c) 0.220 m, 0.219 m, 0.102 m, 0.021 m, 0.02 m, 0.012 m, 0.009 m
   d) 41.1 g, 41.06 g, 40.93 g, 40.81 g, 40.73 g, 40.7 g, 40.07 g

Page 3 — Addition and Subtraction

Q1 a) 1613   **e)** 7271
   b) 5005   **f)** 7279
   c) 78   **g)** 1890
   d) 1038   **h)** 10 440

Q2 a) -5   **e)** 13
   b) -14   **f)** -6
   c) -9   **g)** 0
   d) 7   **h)** 6

Q3 a) 158.3   **e)** 21.5
   b) 70.8   **f)** 204.03
   c) 256.41   **g)** 66.241
   d) 1210.146   **h)** 0.831

Q4 a) 456.1159   **d)** 0.00144
   b) 102.283   **e)** 3.59
   c) 714.5952   **f)** 85.81

Page 4 — Multiplying Without a Calculator

Q1 a) 510   **d)** 160000
   b) 3200   **e)** 760
   c) 1400   **f)** 54.87

Q2 a) 2021   **f)** 53.75
   b) 15730   **g)** -10.22
   c) 245861   **h)** -13.133
   d) 245014   **i)** 5289
   e) 79.81   **j)** 25.9896

Q3 a) 20.097   **e)** 200.97
   b) 2009.7   **f)** 200.97
   c) 200.97   **g)** 200.97
   d) 2009.7   **h)** 2.0097

Q4 a) 460
   b) £12 270
   c) £3830

Page 5 — Dividing Without a Calculator

Q1 a) 35   **d)** 0.2
   b) 15   **e)** 0.16
   c) 190   **f)** 4.1036

Q2 a) 51   **e)** 11
   b) 248   **f)** 23
   c) 169   **g)** 29
   d) 12   **h)** 816

Q3 a) 246 rem 1
   b) 255 rem 2
   c) 61 rem 5
   d) 32 rem 7
   e) 12 rem 3
   f) 21 rem 20
   g) 21 rem 2
   h) 51 rem 10

Q4 12 cages

Q5 6 bags

Q6 122 days

Page 6 — Special Types of Number

Q1 a) Even
   b) Odd
   c) Square
   d) Cube

Q2 4, 16, 36, 64

Q3 a) 16   **c)** 64
   b) 15   **d)** 512

Q4 a) 25, 15, 21, 11, 27
   b) 25, 16, 64
   c) 20, 16, 64
   d) 27, 64

Q5 a) 2, 4, 6, 8, 10, 12, 14, 16, 18, 20
   b) 1, 3, 5, 7, 9, 11, 13, 15, 17, 19, 21, 23, 25, 27, 29
   c) 1, 4, 9, 16, 25
   d) 1, 8, 27, 64, 125, 216, 343, 512

Q6 a) 3
   b) 7.437, 3, $\frac{1}{2}$
   c) $\sqrt{2}$, $\sqrt{3}$, π
   d) 7.437, 3, $\sqrt{2}$, $\frac{1}{2}$, π, $\sqrt{3}$

Page 7 — Multiples and Factors

Q1 a) 0 or 5.
   b) Always a multiple of 3.
   c) Last two digits form a 2 digit number that divides by 4.

Q2 Yes.

Q3 24

Answers

Q4 a) 6, 12, 18, 24, 30, 36, 42, 48, 54, 60, 66, 72
8, 16, 24, 32, 40, 48, 56, 64, 72, 80
b) 24, 48, 72
c) 24

Q5 a) 1, 2, 3, 4, 6, 12
1, 2, 3, 6, 9, 18
1, 2, 3, 4, 6, 8, 12, 24
1, 2, 3, 5, 6, 10, 15, 30
b) 1, 2, 3, 6
c) 6

Q6 a) 16
b) 15
c) 12

Q7 1, 2, 3, 4, 5, 6, 10, 12, 15, 20, 25, 30, 50, 60, 75, 100, 150, 300

Q8 200 seconds' time

Q9 12:07

Page 8 — Primes and Prime Factors

Q1

Q2 71, 83, 107 are prime.

Q3 No

Q4 $90 = 2 \times 3^2 \times 5$
$120 = 2^3 \times 3 \times 5$
$140 = 2^2 \times 5 \times 7$
$180 = 2^2 \times 3^2 \times 5$
$210 = 2 \times 3 \times 5 \times 7$
$864 = 2^5 \times 3^3$
$1000 = 2^3 \times 5^3$

Q5 $504 = 2^3 \times 3^2 \times 7$

Q6 $10 = 3 + 7$
$20 = 13 + 7$ or $17 + 3$
$30 = 23 + 7$ or $17 + 13$ or $11 + 19$

Q7 a) 7
b) 3^2
c) 47
d) $3 \times 5 \times 7$
e) $2^3 \times 3^4$
f) $2^2 \times 5 \times 11$
g) 5×3^4
h) $2^6 \times 3^4 \times 5$

Q8 35: 5, 7
784: 2, 4, 7, 8, 14, 16, 28, 49, 56, 98, 112, 196, 392
20: 2, 4, 5, 10

Page 9 — Fractions, Decimals and Percentages

Q1 a) 28%
b) 57%
c) 87.5%
d) 47.25%
e) 4%
f) 4.5%

Q2 a) 0.35
b) 0.358
c) 0.7
d) 0.07
e) 0.007
f) 0.055

Q3 a) 87.5%
b) 31.25%
c) 32.5%
d) 85%
e) 56%
f) 83.75%

Q4 a) $0.222 = 22.2\%$
b) $0.867 = 86.7\%$
c) $0.389 = 38.9\%$
d) $0.364 = 36.4\%$
e) $0.583 = 58.3\%$
f) $0.385 = 38.5\%$

Q5 a) $\frac{1}{8}$
b) $\frac{3}{8}$
c) $\frac{5}{8}$
d) $\frac{7}{8}$
e) $\frac{3}{40}$
f) $\frac{7}{40}$

Q6 a) English: 74%,
History: 70%,
Maths: 83.5%,
Basket Weaving: 65%
b) Best result — Maths
Worst result — Basket Weaving

Q7 Jamila got 72%, Diana got 80%
So Diana got the higher mark.

Page 10 — Fractions

Q1 a) $\frac{4}{5} > \frac{3}{4}$
b) $\frac{2}{3} > \frac{5}{8}$
c) $\frac{2}{5} > \frac{1}{3}$
d) $\frac{7}{10} > \frac{13}{20}$

Q2 a) $\frac{2}{5}$
b) 1
c) $\frac{1}{2}$
d) 2
e) $2\frac{1}{5}$ or $\frac{11}{5}$
f) $3\frac{1}{3}$ or $\frac{10}{3}$
g) $5\frac{5}{6}$ or $\frac{35}{6}$
h) $10\frac{5}{12}$ or $\frac{125}{12}$

Q3 a) $\frac{3}{4}$
b) $\frac{2}{5}$
c) $\frac{1}{3}$
d) $\frac{3}{5}$
e) $1\frac{3}{8}$ or $\frac{11}{8}$
f) $1\frac{1}{2}$ or $\frac{3}{2}$
g) $2\frac{1}{4}$ or $\frac{9}{4}$
h) $1\frac{1}{5}$ or $\frac{6}{5}$

Q4 a) £16
b) 6 kg
c) 4000 people
d) 45 days
e) 60°
f) £2700

Q5 a) $\frac{9}{16}$
b) $\frac{1}{20}$
c) $3\frac{3}{8} = \frac{27}{8}$
d) $7\frac{11}{15} = \frac{116}{15}$

Q6 a) $1\frac{3}{5} = \frac{8}{5}$
b) $2\frac{4}{5} = \frac{14}{5}$
c) $1\frac{5}{9} = \frac{14}{9}$
d) 2

Q7 a) 360 hectares
b) $\frac{3}{5}$
c) 180 hectares

Pages 11-12 — Percentage Basics

Q1 a) 35p
b) 36p
c) £4.50
d) £4.80
e) 88p
f) 70p
g) £3.00
h) £300
i) £625

Q2 a) £512
b) £9.76
c) 148.5 square miles
d) 5980 people
e) 56 lizards
f) 405 cars

Q3 a) Out of 44500 voters in the town, 14240 voted for the Conservatives.
b) 630 out of 3500 cars stopped had defects.
c) 67.5 grams of the cake's weight of 450 grams is butter.
d) 247 of the 1300 rare birds found were diseased.

Answers

Q4 a) 22 children
  b) 63 grams
  c) 90 lorries
  d) 204 insects
  e) 42
  f) 325
  g) £2.25
  h) £105
  i) 361.8
  j) 2410 grams

Q5 Yes, always true.

Q6 a) 30% of people in Darkley believe in ghosts.
  b) 80% of people are against annoying ringtones on buses.
  c) 12.5% of workers are off sick at present.
  d) Only 15% of children thought there should be more homework.

Q7 5%

Q8 11.3%

Q9 80.8%

Q10 No (only 26%)

Q11 a) £1500
  b) 10%
  c) 83 1/3 % or 83.3% to 1 d.p.

Page 13 — Rounding Numbers

Q1 a) 4.7
  b) 8.9
  c) 6.8
  d) 19.5
  e) 11.8
  f) 20.9

Q2 a) 4.76
  b) 5.09
  c) 17.09
  d) 12.99
  e) 14.99
  f) 17.10

Q3 a) 4.869 kg
  b) 1.009 kg
  c) 1.010 kg
  d) 2.071 kg
  e) 3.061 kg
  f) 0.004 kg

Q4 a) 12.8°
  b) 12.9°
  c) 27.0°
  d) 25.0°
  e) 57.8°
  f) 57.9°

Q5 a) 5.8 km
  b) 9.0 km
  c) 8.5 km
  d) 8.4 km
  e) 17.7 km
  f) 17.7 km

Q6 a) 6760
  b) 6770
  c) 6770
  d) 2010
  e) 2010
  f) 2000

Q7 a) 0.35
  b) 0.036
  c) 0.0057
  d) 4.0
  e) 0.036
  f) 1.0

Q8 a) Bigtown: 369000
  b) Shortville: 102000
  c) Middlethorpe: 191000
  d) Littlewich: 130000
  e) Megaborough: 480000
  f) Port Average: 157000

Page 14 — Rounding Errors and Estimating

Q1 a) -0.2
  b) -0.4
  c) -0.21
  d) 0.44
  e) 112
  f) -3012

Q2 a) $115 \leq x < 125$
  b) $250 \leq x < 350$
  c) $10.15 \leq x < 10.25$
  d) $7.45 \leq x < 7.55$
  e) $8750 \leq x < 8850$
  f) $1005 \leq x < 1015$

Q3 a) i) 36 **ii)** 33.9
  b) i) 20 **ii)** 20.7
  c) i) 63 **ii)** 64.0
  d) i) 6 **ii)** 6.48
  e) i) 1 **ii)** 1.06
  f) i) 50 **ii)** 47.7

Q4 a) i) 2 **ii)** 2.292
  b) i) 0.2 **ii)** 0.214
  c) i) 0.3 **ii)** 0.307
  d) i) 1.2 **ii)** 1.625
  e) i) 1.2 **ii)** 1.277
  f) i) 2.4 **ii)** 2.294

Q5 a) 80 km/h
  b) 79.6 km/h

Page 15 — Powers

Q1 a) 32 **g)** 100 000
  b) 27 **h)** 1 000 000
  c) 16 **i)** 512
  d) 256 **j)** 216
  e) 81 **k)** 343
  f) 125 **l)** 1 000 000

Q2 a) $4^5 = 1024$
  b) $2^8 = 256$
  c) $3^9 = 19\ 683$
  d) $2^{11} = 2048$
  e) $3^2 = 9$
  f) $10^3 = 1000$
  g) $5^2 = 25$
  h) $7^1 = 7$
  i) $4^3 = 64$

Q3 a) a^3 **g)** x^3y
  b) $2a^3$ **h)** x^2y^3
  c) $6x^3$ **i)** $5abc$
  d) $20y^2$ **j)** $12x^2y$
  e) xy **k)** $8xy$
  f) xyz **l)** $10j^2k^2$

Q4 a) x^6 **d)** b^3
  b) y^3 **e)** r^4
  c) a^3 **f)** y^3

Q5 a) $60a^2$ **f)** $21m^2n$
  b) $36x^3$ **g)** x^4
  c) $20v^3$ **h)** y^{12}
  d) $6a^5$ **i)** x^{-6}
  e) $24p^6$

Q6 a) $2x$ **d)** $4k^3$
  b) $5a^3$ **e)** $\dfrac{3x^5}{2y^5}$
  c) $3b$

Q7 a) 10^{-2} **d)** a^{-4}
  b) x^{-2} **e)** $5a^{-4}$
  c) 10^{-4}

Page 16 — Square Roots and Cube Roots

Q1 a) 7.07 **d)** 3.87
  b) 4.47 **e)** 2.65
  c) 8.06 **f)** 8.49

Q2 a) 4.3 **d)** 2.9
  b) 5.3 **e)** 4.0
  c) 1.5 **f)** 2.2

Q3 a) 7 and -7
  b) 16 and -16
  c) 9.5 and -9.5
  d) 9.3 and -9.3

Q4 a) 9 and -9
  b) 5 and -5
  c) 4 and -4
  d) 10 and -10
  e) 2 and -2
  f) 6 and -6

Answers

Q5 a) 5 d) 3
b) 4 e) 1
c) 2 f) 0

Q6 a) $4x$ f) a^2
b) $5a$ g) $3a$
c) $10m$ h) $4ab$
d) $8ab$ i) $10a^2$
e) $4abc$ j) a^2

Page 17 — Standard Form

Q1 a) 5×10^3 f) 3×10^7
b) 9×10^3 g) 3×10^8
c) 9×10^4 h) 8×10^9
d) 2×10^5 i) 1×10^{10}
e) 3×10^6

Q2 a) 5×10^6
b) 5.8×10^6
c) 5.85×10^6
d) 6×10^6
e) 6.7×10^6
f) 6.75×10^6

Q3 a) 4000 f) 64 000
b) 4300 g) 64 200
c) 4350 h) 64 250
d) 435 200 i) 64 258
e) 60 000

Q4 a) 3.5×10^6
b) 1.6×10^5
c) 4.5×10^7
d) 1.27×10^8
e) 5.85×10^5
f) 7.28×10^{10}
g) 3×10^4
h) 8.5×10^5
i) 3×10^3

Q5 a) 6×10^9
b) 3×10^{10}
c) 1.2×10^{14}
d) 1.25×10^{10}

Q6 a) 3×10^5
b) 4×10^8
c) 3.5×10^6
d) 2×10^2

Q7 a) 4×10^{-4}
b) 2×10^{-2}
c) 2.5×10^{-2}
d) 5×10^{-4}
e) 5.2×10^{-4}
f) 5.27×10^{-4}

Q8 1.76×10^3, 2.31×10^3, 2450

Q9 6.5×10^{-5}, 1.6×10^{-4}, 0.0078

Section 2 — Algebra

Page 18 — Algebra — Simplifying Terms

Q1 a) $2y + 2z$ or $2(y + z)$
b) $8a$
c) $6w - 2v$
d) $5c - 5d$ or $5(c - d)$
e) $2u - 2t + 10$ or $2(u - t + 5)$
f) $4f - 2g + 1$
g) $5r - 4s - 3$
h) $-h - 3j - 1$

Q2 a) a^2 e) $2d^2f^2$
b) b^3 f) g^4
c) c^4 g) $2h^2g^2$
d) a^2b h) $12k^4$

Q3 a) z e) $5m$
b) y^3 f) $2n^5$
c) x^4 g) 1
d) $2k$ h) 1

Q4 $p + p = 2p$
$a - b + a = 2a - b$
$r + s - r = s$
$(t + t) \div 4u = \dfrac{t}{2u}$
$v + 2 + v - 5 = 2v - 3$
$a \times b = ba$
$\dfrac{1}{2} \times w \times x = \dfrac{wx}{2}$
$3 + y - 7 - y + 4 = 0$

Page 19 — Algebra — Expanding and Factorising

Q1 a) $2p + 6$ e) $6s + 3$
b) $3q - 9$ f) $12s - 28$
c) $16 + 4r$ g) $50 + 15t$
d) $10 - 5r$ h) $-12 - 12u$

Q2 a) $3a + 3b$
b) $3m + 6n + 15k$
c) $12x - 18y$
d) $50x - 40y + 60z$
e) $-2x - 2y$
f) $-4c + 10d$
g) $24a + 32b - 48$
h) $2ab + 6ac$
i) $-3mx - 3my$
j) $-3m^2 - 3mn$
k) $4hl - 4h^2$
l) $6y^3m + 6yn$

Q3 a) $2(a + 2)$ f) $3(x + 2y)$
b) $3(2b + 3)$ g) $a(x + y)$
c) $3(c - 2)$ h) $5(2a + 3b)$
d) $4(2 - d)$ i) $2(2x - y)$
e) $2(x + y)$ j) $3x(2 - 3z)$

Q4 a) $7p(p - 2q)$ e) $6b(2x - y + 4z)$
b) $a(ap - q)$ f) $a(1 - 4a)$
c) $3y(xy + y + z)$ g) $n(n + 5)$
d) $4a(x + 2y)$ h) $x(x - 1)$

Q5 a) $x^2 + 6x + 8$
b) $x^2 + 12x + 36$
c) $x^2 + 2x - 15$
d) $x^2 - 6x + 9$
e) $2a^2 + 9a + 9$
f) $a^2 + a - 12$
g) $m^2 - 5m + 6$
h) $2m^2 + 7m + 5$
i) $3y^2 - 13y - 10$
j) $16x^2 + 24x + 9$
k) $2k^3 + 11k^2 + 13k + 4$
l) $3y^3 + 5y^2 - 6y - 8$

Page 20 — Solving Equations

Q1 a) $x = 5$ h) $x = -8$
b) $x = 4$ i) $x = 44$
c) $x = 8$ j) $x = 21$
d) $x = 4$ k) $x = 40$
e) $x = 19$ l) $x = 10$
f) $x = -2$ m) $x = 3$
g) $x = -9$ n) $x = 0.5$

Q2 a) $x = 4$ d) $x = 4$
b) $x = 20$ e) $x = 10$
c) $x = 10$ f) $x = 100$

Q3 a) $x = 10$ f) $x = 15$
b) $x = 42$ g) $x = 30$
c) $x = 20$ h) $x = 100$
d) $x = 5$ i) $x = 7$
e) $x = 3$ j) $x = 100$

Q4 a) $x = 0.75$
b) $x = -1$
c) $x = -6$
d) $x = -1$
e) $x = 0.44$ (2 d.p.)
f) $x = 1.18$ (2 d.p.)

Page 21 — Using Formulas

Q1 a) 12 b) 50

Q2 a) 240 b) 0.5

Q3 a) 20.58 cm^3
b) 3.125 m
c) 15 mm

Q4 a) 22.36 cm^2 b) 8 m

Q5 a) £260
b) £110
c) 75

Q6 a) 570
b) 200
c) 10

Q7 a) 5.5
b) 2
c) 0.5

Answers

Page 22 — Expressions and Formulas from Words

Q1 a) $x + 20°$
  b) $5n$
  c) $y - 5$
  d) $k - 2$ kg
  e) $80L$ metres
  f) $v + 15$ km/h
  g) $0.25y$ litres

Q2 $C = 15 + 10h$

Q3 $c = 1.2m + 1.99$

Q4 $T = 50n$

Q5 $C = 14.5x + 7.6y$

Page 23 — Equations and Formulas from Words

Q1 a) $£b = 0.08n + 15$
  b) £16.60
  c) £24.60
  d) £19.80

Q2 a) $£x = 0.5d + 0.25e$, where
    d = number of daytime calls, and
    e = number of evening calls
  b) £12.50
  c) £27.50
  d) 92
  e) 0

Q3 a) $b + 22$
  b) $(b + 22) + b = 460$
    $b = 219$
  c) 241

Q4 a) $x = 59°$
  b) $x = 85°$
  c) $x = 30°$
  d) $x = 35°$

Page 24 — Rearranging Formulas

Q1 a) 14   **b)** $x = b - a$
Q2 a) 18   **b)** $x = b + a$
Q3 a) 6   **b)** $x = \frac{q}{p}$
Q4 a) 27   **b)** $x = nm$
Q5 a) 7   **b)** $x = c - d$
Q6 a) 19   **b)** $x = h + k$
Q7 a) 5   **b)** $x = \frac{u}{v}$
Q8 a) 60   **b)** $x = ab$
Q9 a) 4   **b)** $x = \frac{c - b}{u}$
Q10 a) 8   **b)** $x = \frac{r + q}{p}$
Q11 a) 4   **b)** $x = \frac{a + c}{b}$
Q12 a) 9   **b)** $x = m(p - n)$
Q13 a) 105 **b)** $x = m(l + n)$

Q14 a) 33   **b)** $x = s(t - r)$
Q15 a) 2   **b)** $x = \frac{z - m}{n}$
Q16 a) 200   **b)** $x = k(e + r)$
Q17 a) -5   **b)** $x = b - a$
Q18 a) 3   **b)** $x = \frac{q - p}{r}$

Q19 a) $T = \frac{v}{a}$
  b) $T = \frac{d}{ax}$
  c) $T = \frac{p}{5rs}$
  d) $S = \frac{p}{5rt}$
  e) $Q = \frac{m}{p}$
  f) $Q = \frac{m}{p^2}$
  g) $R = \frac{c}{2p}$
  h) $H = \frac{v}{lb}$
  i) $B = 100a$
  j) $B = \frac{100a}{c}$
  k) $R = \frac{100l}{pt}$
  l) $B = 3a$

Q20 a) $B = c - a$
  b) $K = (l - 2a)^2$
  c) $C = b - d^2$
  d) $C = 3b - 2d$
  e) $A = \frac{n^2 - m}{3}$
  f) $P = 2n - 7$
  g) $D = 5 - 3t$
  h) $X = 4(b - a)$ or $X = 4b - 4a$
  i) $M = n(h - k)$ or $M = hn - kn$

Pages 25-26 — Number Patterns and Sequences

Q1 a) 4, 7, 10, 13, 16

  b) 3, 5, 7, 9, 11

  c) 12, 19, 26, 33

Q2

n	1	2	3	4	5	6	7	8
t	3	5	7	9	11	13	15	17

Q3 a) Add 2, Arithmetic; 18, 20, 22
  b) Divide by 2, Geometric; 28, 14, 7
  c) Add 3, Arithmetic; 15, 18, 21
  d) Multiply by 2, Geometric; 48, 96, 192
  e) Subtract 2, Arithmetic; 0, -2, -4
  f) Add 3, Arithmetic; -1, 2, 5
  g) Subtract 3, Arithmetic; -17, -20, -23
  h) Multiply by 3, Geometric; 162, 486, 1458

Q4

n	1	2	3	4	5	6	7	8
t	1	4	7	10	13	16	19	22

Q5 a) 6, 11, 16, 21, 26, 31, 36, 41
  b) 5 5 5 5 5 5 5
  c) $t = 5n + 1$
  d) 101

Q6 a) 7, 9, 11, 13, 15, 17, 19, 21
  b) 2 2 2 2 2 2 2
  c) nth term $= 2n + 5$

Q7 a) 2, 9, 16, 23, 30, 37, 44, 51
  b) nth term $= 7n - 5$
  c) Yes

Q8 a) 4, 6.5, 9, 11.5, 14, 16.5, 19, 21.5
  b) nth term $= 2.5n + 1.5$
  c) 126.5

Q9 a) 5, 2, -1, -4, -7
  b) -3
  c) nth term $= -3n + 23$
  d) No

Q10 a) nth term $= 7n - 4$
  b) nth term $= 4n + 1$
  c) nth term $= -3n + 17$
  d) nth term $= -5n + 32$

Page 27 — Inequalities

Q1 a) 1, 2
  b) -6, -5, -4, -3, -2, -1
  c) 1, 2, 3, 4
  d) -3, -2, -1

Q2 a) 3,4,5,6
  b) 2,3,4,5,6,7
  c) 3,4,5,6,7
  d) 21,22,23,24,25
  e) 9
  f) -1,0,1,2,3
  g) -2,-1,0,1
  h) -7,-6,-5,-4,-3

Answers

Q3 a)
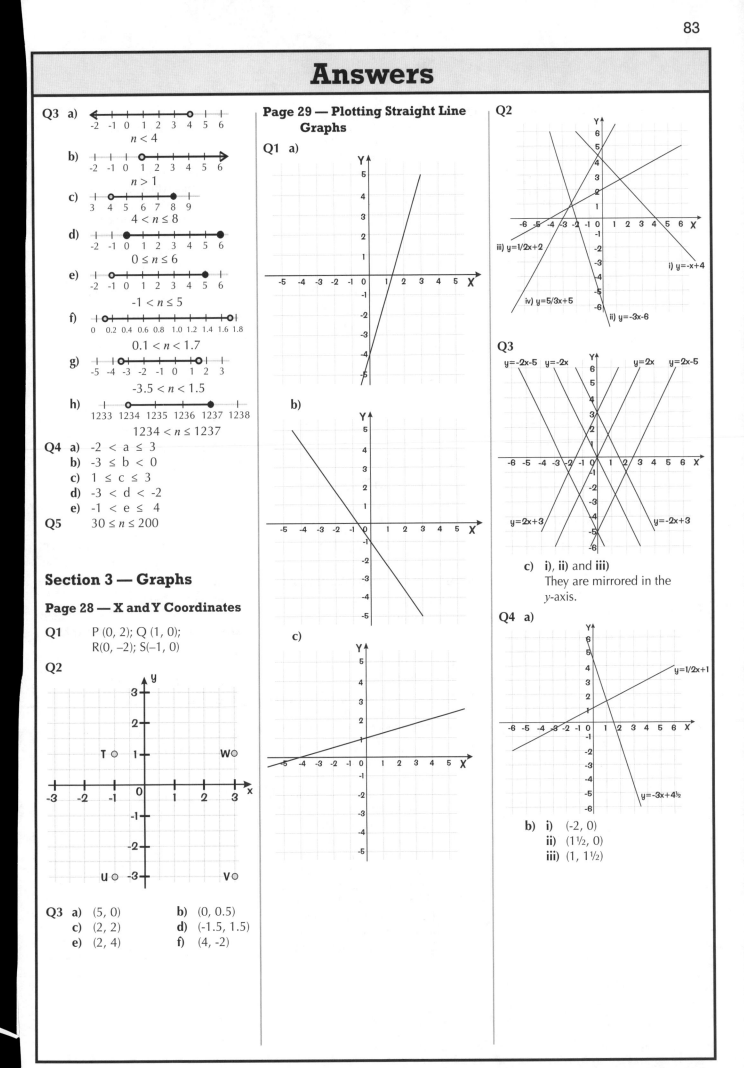
$n < 4$

b)
$n > 1$

c)
$4 < n \le 8$

d)
$0 \le n \le 6$

e)
$-1 < n \le 5$

f)
$0.1 < n < 1.7$

g)
$-3.5 < n < 1.5$

h)
$1234 < n \le 1237$

Q4 a) $-2 < a \le 3$
  b) $-3 \le b < 0$
  c) $1 \le c \le 3$
  d) $-3 < d < -2$
  e) $-1 < e \le 4$

Q5 $30 \le n \le 200$

Section 3 — Graphs

Page 28 — X and Y Coordinates

Q1 P (0, 2); Q (1, 0);
  R(0, –2); S(–1, 0)

Q2

Q3 a) (5, 0)   **b)** (0, 0.5)
  c) (2, 2)   **d)** (-1.5, 1.5)
  e) (2, 4)   **f)** (4, -2)

Page 29 — Plotting Straight Line Graphs

Q1 a)

b)

c)

Q2

iii) y=1/2x+2
i) y=-x+4
iv) y=5/3x+5
ii) y=-3x-6

Q3

y=-2x-5 y=-2x   y=2x   y=2x-5
y=2x+3   y=-2x+3

c) i), ii) and **iii)**
They are mirrored in the
y-axis.

Q4 a)

y=1/2x+1
y=-3x+4½

b) i) (-2, 0)
  ii) (1½, 0)
  iii) (1, 1½)

Answers

Pages 30-31 — Gradients and y = mx + c

Q1

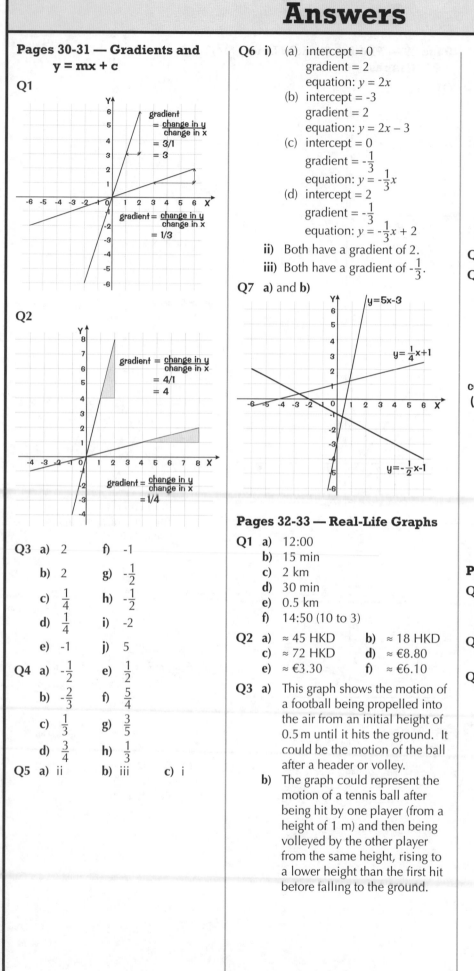

Q2

Q3
a) 2
b) 2
c) $\frac{1}{4}$
d) $\frac{1}{4}$
e) -1
f) -1
g) $-\frac{1}{2}$
h) $-\frac{1}{2}$
i) -2
j) 5

Q4
a) $-\frac{1}{2}$
b) $-\frac{2}{3}$
c) $\frac{1}{3}$
d) $\frac{3}{4}$
e) $\frac{1}{2}$
f) $\frac{5}{4}$
g) $\frac{3}{5}$
h) $\frac{1}{3}$

Q5 a) ii   b) iii   c) i

Q6 i)
(a) intercept = 0
gradient = 2
equation: $y = 2x$
(b) intercept = -3
gradient = 2
equation: $y = 2x - 3$
(c) intercept = 0
gradient = $-\frac{1}{3}$
equation: $y = -\frac{1}{3}x$
(d) intercept = 2
gradient = $-\frac{1}{3}$
equation: $y = -\frac{1}{3}x + 2$

ii) Both have a gradient of 2.
iii) Both have a gradient of $-\frac{1}{3}$.

Q7 a) and b)

Pages 32-33 — Real-Life Graphs

Q1
a) 12:00
b) 15 min
c) 2 km
d) 30 min
e) 0.5 km
f) 14:50 (10 to 3)

Q2
a) ≈ 45 HKD   b) ≈ 18 HKD
c) ≈ 72 HKD   d) ≈ €8.80
e) ≈ €3.30   f) ≈ €6.10

Q3
a) This graph shows the motion of a football being propelled into the air from an initial height of 0.5 m until it hits the ground. It could be the motion of the ball after a header or volley.
b) The graph could represent the motion of a tennis ball after being hit by one player (from a height of 1 m) and then being volleyed by the other player from the same height, rising to a lower height than the first hit before falling to the ground.

c) The graph shows a car accelerating jerkily to about 15 mph, then slowing to a halt. The car then accelerates again up to a speed of 30 mph, which is maintained for a period of time, before it slows again to a halt. The wiggles in the car's acceleration could be the driver changing the pressure on the accelerator, applying the brakes or changing gear.

Q4   1) A, 2) D, 3) B, 4) C

Q5
a) £20
b) £27
c)

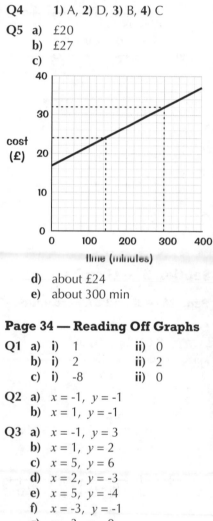

d) about £24
e) about 300 min

Page 34 — Reading Off Graphs

Q1
a) i) 1   ii) 0
b) i) 2   ii) 2
c) i) -8   ii) 0

Q2
a) $x = -1$, $y = -1$
b) $x = 1$, $y = -1$

Q3
a) $x = -1$, $y = 3$
b) $x = 1$, $y = 2$
c) $x = 5$, $y = 6$
d) $x = 2$, $y = -3$
e) $x = 5$, $y = -4$
f) $x = -3$, $y = -1$
g) $x = 3$, $y = 0$
h) $x = -1$, $y = -2$

Answers

Page 35 — Quadratic Graphs

Q1

x	-4	-3	-2	-1	0	1	2	3	4	5
$y = x^2$	16	9	4	1	0	1	4	9	16	25

a) and **b)**

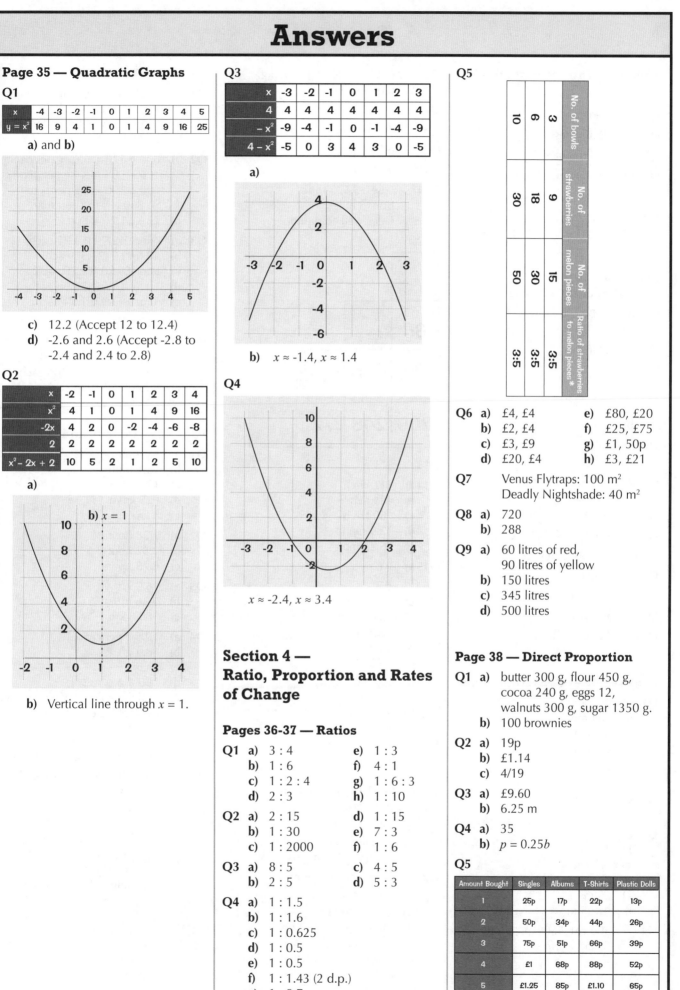

c) 12.2 (Accept 12 to 12.4)
d) -2.6 and 2.6 (Accept -2.8 to -2.4 and 2.4 to 2.8)

Q2

x	-2	-1	0	1	2	3	4
x^2	4	1	0	1	4	9	16
-2x	4	2	0	-2	-4	-6	-8
2	2	2	2	2	2	2	2
$x^2 - 2x + 2$	10	5	2	1	2	5	10

a)

b) $x = 1$

b) Vertical line through $x = 1$.

Q3

x	-3	-2	-1	0	1	2	3
4	4	4	4	4	4	4	4
$-x^2$	-9	-4	-1	0	-1	-4	-9
$4 - x^2$	-5	0	3	4	3	0	-5

a)

b) $x \approx -1.4$, $x \approx 1.4$

Q4

$x \approx -2.4$, $x \approx 3.4$

Section 4 —
Ratio, Proportion and Rates of Change

Pages 36-37 — Ratios

Q1
a) 3 : 4
b) 1 : 6
c) 1 : 2 : 4
d) 2 : 3
e) 1 : 3
f) 4 : 1
g) 1 : 6 : 3
h) 1 : 10

Q2
a) 2 : 15
b) 1 : 30
c) 1 : 2000
d) 1 : 15
e) 7 : 3
f) 1 : 6

Q3
a) 8 : 5
b) 2 : 5
c) 4 : 5
d) 5 : 3

Q4
a) 1 : 1.5
b) 1 : 1.6
c) 1 : 0.625
d) 1 : 0.5
e) 1 : 0.5
f) 1 : 1.43 (2 d.p.)
g) 1 : 0.7
h) 1 : 0.17 (2 d.p.)

Q5

No. of bowls	No. of strawberries	No. of melon pieces	Ratio of strawberries to melon pieces *
3	9	15	3:5
6	18	30	3:5
10	30	50	3:5

Q6
a) £4, £4
b) £2, £4
c) £3, £9
d) £20, £4
e) £80, £20
f) £25, £75
g) £1, 50p
h) £3, £21

Q7 Venus Flytraps: 100 m²
Deadly Nightshade: 40 m²

Q8
a) 720
b) 288

Q9
a) 60 litres of red, 90 litres of yellow
b) 150 litres
c) 345 litres
d) 500 litres

Page 38 — Direct Proportion

Q1
a) butter 300 g, flour 450 g, cocoa 240 g, eggs 12, walnuts 300 g, sugar 1350 g.
b) 100 brownies

Q2
a) 19p
b) £1.14
c) 4/19

Q3
a) £9.60
b) 6.25 m

Q4
a) 35
b) $p = 0.25b$

Q5

Amount Bought	Singles	Albums	T-Shirts	Plastic Dolls
1	25p	17p	22p	13p
2	50p	34p	44p	26p
3	75p	51p	66p	39p
4	£1	68p	88p	52p
5	£1.25	85p	£1.10	65p
6	£1.50	£1.02	£1.32	78p

Answers

Page 39 — Inverse Proportion

Q1 1.5 minutes

Q2 9 hours

Q3 25 minutes

Q4 22.5 minutes

Q5 a) 30 minutes
    b) $t = \dfrac{270}{x}$

Q6 a) 2 months
    b) $m = \dfrac{30}{b}$

Pages 40-41 —
Percentage Change

Q1 a) £9.20
    b) £55.20

Q2 £248.04

Q3 16933

Q4 387

Q5 a) £21
    b) £413

Q6 a) 26%
    b) £350

Q7 a) 14% gain in value.
    b) Glen's net gain is 10.1%.

Q8 £250

Q9 £56

Q10 700 000 litres

Q11 £1680

Q12 361.5 million square km

Q13 £21.00

Q14 11 seconds

Pages 42-43 — Converting Units

Q1 a) 48 mm
    b) 264 mm
    c) 87.5 mm
    d) 6.3 mm

Q2 a) 7.6 cm
    b) 18.5 cm
    c) 350 cm
    d) 0.05 cm

Q3 a) 1.45 m
    b) 3.5 m
    c) 0.85 m
    d) 0.05 m
    e) 25 m
    f) 0.155 m

Q4 500 cm, 560 cm, 568 cm, 75 cm, 5 cm

Q5 1400 g, 2850 g, 650 g

Q6 0.45 kg, 1.45 kg, 2.45 kg, 0.05 kg, 0.005 kg

Q7 20 glasses

Q8 4.118 tonnes

Q9 67.5 litres

Q10 437 minutes

Q11 Giles: 5.625 miles, Bob: 5 miles
     Giles walked further.

Q12

Aberdeen

168 | Inverness

240 | 280 | Glasgow

200 | 248 | 72 | Edinburgh

Q13 511 g

Q14 a) 112 km/h
    b) 48 km/h
    c) 160 km/h
    d) 22.5 mph
    e) 56.25 mph

Q15 a) 1.6 m^2
    b) Yes

Q16 a) 4 cm^2
    b) $50\,000 \text{ cm}^2$
    c) 0.2 km^2
    d) 0.6 m^2
    e) $40\,000 \text{ m}^2$
    f) $15\,000 \text{ cm}^2$

Q17 a) 10 cm^3
    b) $50\,000 \text{ cm}^3$
    c) 0.3 m^3
    d) $5\,000\,000 \text{ mm}^3$

Page 44 — Maps and Scale Drawings

Q1 a) 2.5 m × 1.5 m
    b) Yes, there is 1.5 m of wall
    c) 1.5 m × 75 cm

Q2 45000 cm = 450 m

Q3 a) 1 900 000 cm
    b) 9.5 : 1 900 000
    c) 1 cm represents 2 km, 1 : 200 000
    d) 16 km

Page 45 — Best Buy

Q1 6 ÷ 250 = 0.024 m per penny
    7 ÷ 260 = 0.0269 m per penny
    So best buy = 7 m of ribbon at £2.60

Q2 a) 800 ÷ 120 = 6.67 g per penny
       400 ÷ 90 = 4.44 g per penny
       600 ÷ 100 = 6 g per penny
       So 800 g is the best buy
    b) £2 – £1.20 = 80p

Q3 a) 4 for £12.99 (£3.25 each)
    b) 40 for £9 (22.5p each)
    c) 5 for £8.50 (£1.70 each)
    d) 10 for £3.99 (39.9p each)
    e) 3 for £3.99 (£1.33 each)

Q4 300 ml for £1.39

Q5 a) 14 jam tarts for £1.40 (10p each)
    b) 10 chews for £3 (30p each)
    c) 6 for £5.95 (99p each)

Page 46 — Density and Speed

Q1 a) 45 mph
    b) 3.25 mph
    c) 125 km/h
    d) 625 km/h
    e) 8 km/s

Q2 a) 1.28 hours
    b) 0.82 hours
    c) 6.25 hours

Q3 a) 462.5 m
    b) 0.36 m
    c) 388.8 km

Q4 a) 10.5 g per cm^3
    b) 88.4 g
    c) 3 cm^3
    d) 166.4 g
    e) Call the volume of ice V. Then
       Mass of water = 1 × 810
                     = 810 g
       Mass of ice = 0.9 × V
       The water and ice will have the same mass, so $0.9V = 810$.
       $V = 810 ÷ 0.9 = 900 \text{ cm}^3$

Q5 a) $\dfrac{\text{Mass of Earth}}{\text{Mass of Moon}} = \dfrac{5.52V}{3.34V}$
       $= 5.52 ÷ 3.34 = 1.7 \text{ (2 sf)}$

    b) $\dfrac{\text{Mass of Earth}}{\text{Mass of Moon}} = \dfrac{5.52 \times 49V}{3.34V}$
       $= 270.48 ÷ 3.34 = 81 \text{ (2 sf)}$

Answers

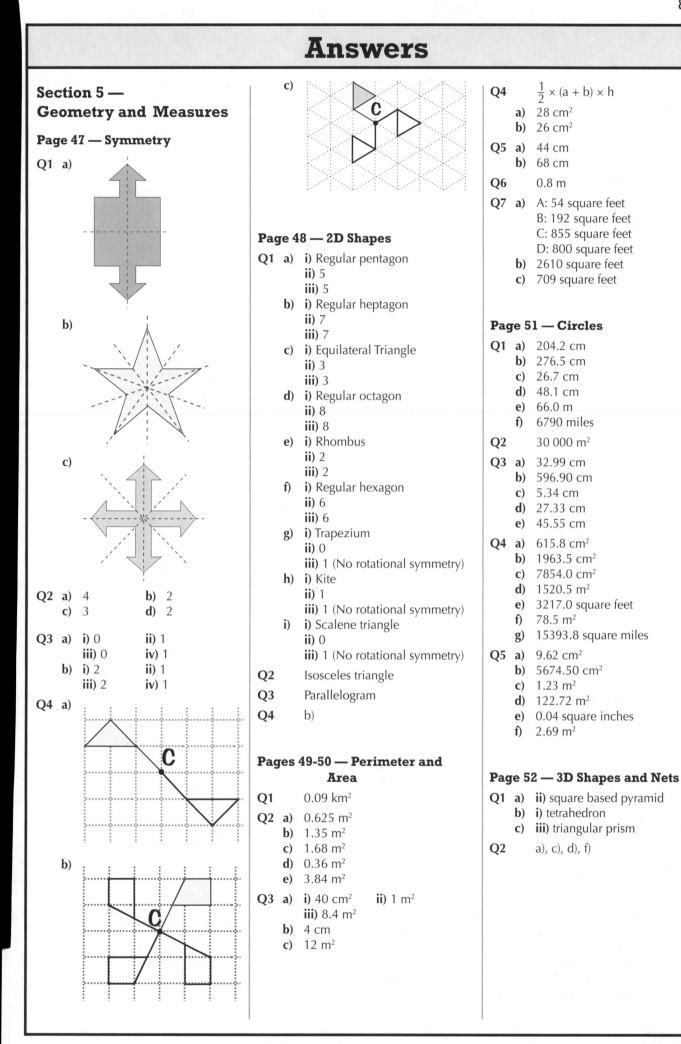

Section 5 —
Geometry and Measures

Page 47 — Symmetry

Q1 a)

b)

c)

Q2 a) 4   **b)** 2
c) 3   **d)** 2

Q3 a) i) 0   **ii)** 1
    iii) 0   **iv)** 1
b) i) 2   **ii)** 1
    iii) 2   **iv)** 1

Q4 a)

b)

c)

Page 48 — 2D Shapes

Q1 a) i) Regular pentagon
    ii) 5
    iii) 5
b) i) Regular heptagon
    ii) 7
    iii) 7
c) i) Equilateral Triangle
    ii) 3
    iii) 3
d) i) Regular octagon
    ii) 8
    iii) 8
e) i) Rhombus
    ii) 2
    iii) 2
f) i) Regular hexagon
    ii) 6
    iii) 6
g) i) Trapezium
    ii) 0
    iii) 1 (No rotational symmetry)
h) i) Kite
    ii) 1
    iii) 1 (No rotational symmetry)
i) i) Scalene triangle
    ii) 0
    iii) 1 (No rotational symmetry)

Q2 Isosceles triangle

Q3 Parallelogram

Q4 b)

Pages 49-50 — Perimeter and Area

Q1 0.09 km²

Q2 a) 0.625 m²
b) 1.35 m²
c) 1.68 m²
d) 0.36 m²
e) 3.84 m²

Q3 a) i) 40 cm²   **ii)** 1 m²
    iii) 8.4 m²
b) 4 cm
c) 12 m²

Q4 $\frac{1}{2} \times (a + b) \times h$
a) 28 cm²
b) 26 cm²

Q5 a) 44 cm
b) 68 cm

Q6 0.8 m

Q7 a) A: 54 square feet
    B: 192 square feet
    C: 855 square feet
    D: 800 square feet
b) 2610 square feet
c) 709 square feet

Page 51 — Circles

Q1 a) 204.2 cm
b) 276.5 cm
c) 26.7 cm
d) 48.1 cm
e) 66.0 m
f) 6790 miles

Q2 30 000 m²

Q3 a) 32.99 cm
b) 596.90 cm
c) 5.34 cm
d) 27.33 cm
e) 45.55 cm

Q4 a) 615.8 cm²
b) 1963.5 cm²
c) 7854.0 cm²
d) 1520.5 m²
e) 3217.0 square feet
f) 78.5 m²
g) 15393.8 square miles

Q5 a) 9.62 cm²
b) 5674.50 cm²
c) 1.23 m²
d) 122.72 m²
e) 0.04 square inches
f) 2.69 m²

Page 52 — 3D Shapes and Nets

Q1 a) ii) square based pyramid
b) i) tetrahedron
c) iii) triangular prism

Q2 a), c), d), f)

Answers

Q3 a)

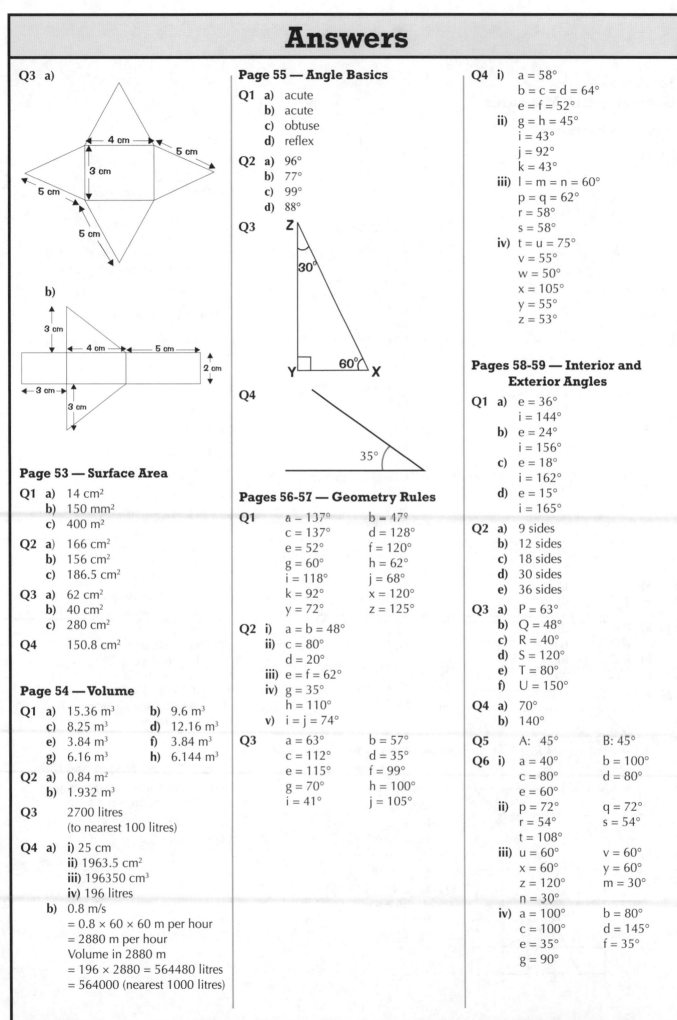

b)

Page 53 — Surface Area

Q1 a) 14 cm²
b) 150 mm²
c) 400 m²

Q2 a) 166 cm²
b) 156 cm²
c) 186.5 cm²

Q3 a) 62 cm²
b) 40 cm²
c) 280 cm²

Q4 150.8 cm²

Page 54 — Volume

Q1 a) 15.36 m³   b) 9.6 m³
c) 8.25 m³   d) 12.16 m³
e) 3.84 m³   f) 3.84 m³
g) 6.16 m³   h) 6.144 m³

Q2 a) 0.84 m²
b) 1.932 m³

Q3 2700 litres
(to nearest 100 litres)

Q4 a) i) 25 cm
ii) 1963.5 cm²
iii) 196350 cm³
iv) 196 litres
b) 0.8 m/s
= 0.8 × 60 × 60 m per hour
= 2880 m per hour
Volume in 2880 m
= 196 × 2880 = 564480 litres
= 564000 (nearest 1000 litres)

Page 55 — Angle Basics

Q1 a) acute
b) acute
c) obtuse
d) reflex

Q2 a) 96°
b) 77°
c) 99°
d) 88°

Q3

Q4

Pages 56-57 — Geometry Rules

Q1
a = 137°	b = 47°
c = 137°	d = 128°
e = 52°	f = 120°
g = 60°	h = 62°
i = 118°	j = 68°
k = 92°	x = 120°
y = 72°	z = 125°

Q2 i) a = b = 48°
ii) c = 80°
d = 20°
iii) e = f = 62°
iv) g = 35°
h = 110°
v) i = j = 74°

Q3
a = 63°	b = 57°
c = 112°	d = 35°
e = 115°	f = 99°
g = 70°	h = 100°
i = 41°	j = 105°

Q4 i) a = 58°
b = c = d = 64°
e = f = 52°
ii) g = h = 45°
i = 43°
j = 92°
k = 43°
iii) l = m = n = 60°
p = q = 62°
r = 58°
s = 58°
iv) t = u = 75°
v = 55°
w = 50°
x = 105°
y = 55°
z = 53°

Pages 58-59 — Interior and Exterior Angles

Q1 a) e = 36°
i = 144°
b) e = 24°
i = 156°
c) e = 18°
i = 162°
d) e = 15°
i = 165°

Q2 a) 9 sides
b) 12 sides
c) 18 sides
d) 30 sides
e) 36 sides

Q3 a) P = 63°
b) Q = 48°
c) R = 40°
d) S = 120°
e) T = 80°
f) U = 150°

Q4 a) 70°
b) 140°

Q5 A: 45°   B: 45°

Q6 i) a = 40°   b = 100°
c = 80°   d = 80°
e = 60°
ii) p = 72°   q = 72°
r = 54°   s = 54°
t = 108°
iii) u = 60°   v = 60°
x = 60°   y = 60°
z = 120°   m = 30°
n = 30°
iv) a = 100°   b = 80°
c = 100°   d = 145°
e = 35°   f = 35°
g = 90°

Answers

Pages 60-61 — Transformations

Q1

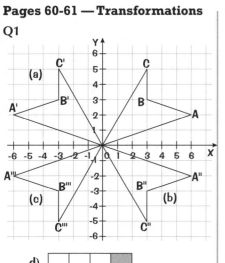

d)

	Original	Reflection (a)	Reflection (b)	Reflection (c)
A (6, 2)	A' (-6, 2)	A" (6, -2)	A''' (-6, -2)
B (3, 3)	B' (-3, 3)	B" (3, -3)	B''' (-3, -3)
C (3, 5)	C' (-3, 5)	C" (3, -5)	C''' (-3, -5)

e) **i)** When reflecting in the x-axis, the sign of the y coordinate changes but the sign of the x coordinate doesn't.

ii) When reflecting in the y-axis, the sign of the x coordinate changes but the sign of the y coordinate doesn't.

Q2

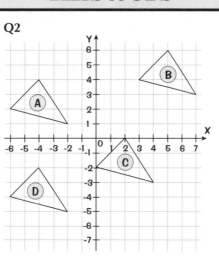

Q3 **a)** Reflection in the y-axis
b) Reflection in the x-axis
c) Rotation 90° anti-clockwise about (0,0)
d) Rotation 90° anti-clockwise about (0,0)
e) Rotation 180° clockwise or anti-clockwise about (0,0)
f) Reflection in the line ML (which is the line y = x)

Q4

	Original	Enlargement ×2	Enlargement ×3
A (3, 0)	A' (6, 0)	A" (9, 0)
B (3, 1)	B' (6, 2)	B" (9, 3)
C (1½, 2)	C' (3, 4)	C" (4½, 6)
D (0, 1)	D' (0, 2)	D" (0, 3)

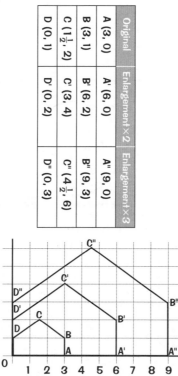

Page 62 — Congruence and Similarity

Q1 A, D   E, B   C, F

Q2 Pair A — yes. The sides of the pale triangle are the same lengths as those of the dark triangle.
Pair B — yes. Two angles and one side are the same on both triangles.

Q3 b), c)

Q4 b), c)

Page 63 — Constructions

Q1 Correctly drawn triangles.

Q2 TUV and WXY are congruent.

Q3 Correctly drawn triangles.

Q4 Accept 2, 2.1, 2.2 cm

Page 64 — Pythagoras' Theorem

Q1 **a)** Less than 90°. $5^2 + 4^2 = 41$
$6.2^2 < 41$
b) More than 90°.
$6.8^2 + 7^2 = 95.24$
$10^2 > 95.24$
c) Right-angle.
$4.5^2 + 6^2 = 56.25$
$56.25 = 7.5^2$

Q2 **a)** 12.0 cm   **b)** 0.206 cm
c) 11.2 cm   **d)** 6.32 cm
e) 9.12 cm   **f)** 0.687 cm
g) 8.00 cm   **h)** 6.00 cm
i) 7.07 cm   **j)** 0.707 cm
k) 9.57 cm

Pages 65-66 — Trigonometry

Q1

Angle a	cos a	sin a	tan a
0°	1	0	0
10°	0.985	0.174	0.176
15°	0.966	0.259	0.268
30°	0.866	0.5	0.577
45°	0.707	0.707	1
60°	0.5	0.866	1.732
80°	0.174	0.985	5.671
88°	0.035	0.999	28.636
90°	0	1	∞

Answers

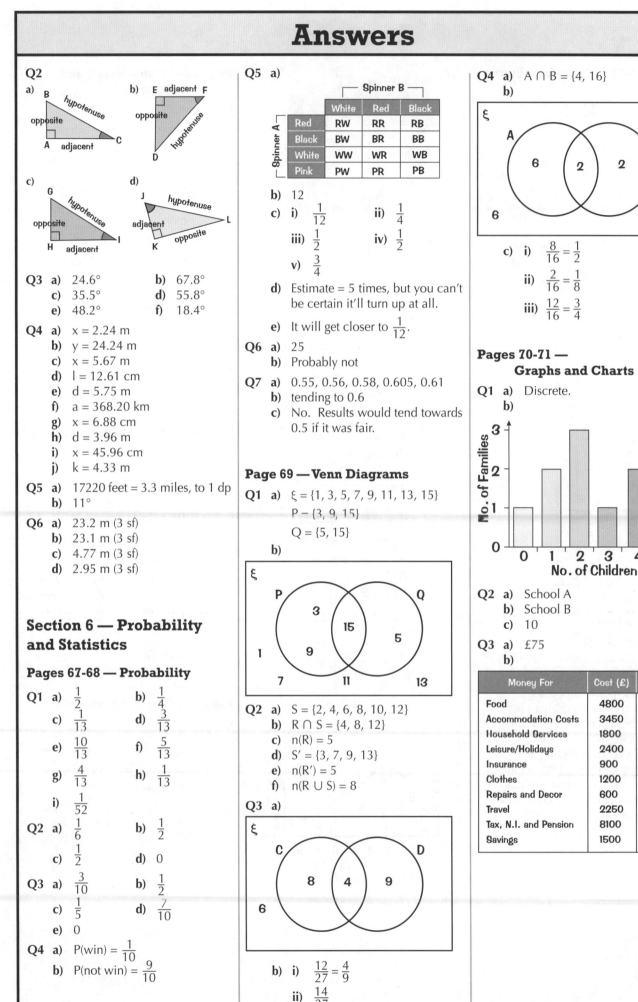

Q2

a) Triangle: B (hypotenuse), opposite at A, adjacent, right angle at A, C

b) E adjacent F, opposite, hypotenuse, D

c) G hypotenuse, opposite, H adjacent I

d) J hypotenuse L, adjacent, opposite, K

Q3 a) 24.6°   b) 67.8°
c) 35.5°   d) 55.8°
e) 48.2°   f) 18.4°

Q4 a) x = 2.24 m
b) y = 24.24 m
c) x = 5.67 m
d) l = 12.61 cm
e) d = 5.75 m
f) a = 368.20 km
g) x = 6.88 cm
h) d = 3.96 m
i) x = 45.96 cm
j) k = 4.33 m

Q5 a) 17220 feet = 3.3 miles, to 1 dp
b) 11°

Q6 a) 23.2 m (3 sf)
b) 23.1 m (3 sf)
c) 4.77 m (3 sf)
d) 2.95 m (3 sf)

Section 6 — Probability and Statistics

Pages 67-68 — Probability

Q1 a) $\frac{1}{2}$ b) $\frac{1}{4}$
c) $\frac{1}{13}$ d) $\frac{3}{13}$
e) $\frac{10}{13}$ f) $\frac{5}{13}$
g) $\frac{4}{13}$ h) $\frac{1}{13}$
i) $\frac{1}{52}$

Q2 a) $\frac{1}{6}$ b) $\frac{1}{2}$
c) $\frac{1}{2}$ d) 0

Q3 a) $\frac{3}{10}$ b) $\frac{1}{2}$
c) $\frac{1}{5}$ d) $\frac{7}{10}$
e) 0

Q4 a) P(win) = $\frac{1}{10}$
b) P(not win) = $\frac{9}{10}$

Q5 a)

	Spinner B		
	White	Red	Black
Red	RW	RR	RB
Black	BW	BR	BB
White	WW	WR	WB
Pink	PW	PR	PB

(Spinner A down the left: Red, Black, White, Pink)

b) 12
c) i) $\frac{1}{12}$ ii) $\frac{1}{4}$
   iii) $\frac{1}{2}$ iv) $\frac{1}{2}$
   v) $\frac{3}{4}$

d) Estimate = 5 times, but you can't be certain it'll turn up at all.

e) It will get closer to $\frac{1}{12}$.

Q6 a) 25
b) Probably not

Q7 a) 0.55, 0.56, 0.58, 0.605, 0.61
b) tending to 0.6
c) No. Results would tend towards 0.5 if it was fair.

Page 69 — Venn Diagrams

Q1 a) ξ = {1, 3, 5, 7, 9, 11, 13, 15}
P = {3, 9, 15}
Q = {5, 15}

b) [Venn diagram: ξ, P and Q overlapping. P region: 3, 9. Intersection: 15. Q region: 5. Outside: 1, 7, 11, 13]

Q2 a) S = {2, 4, 6, 8, 10, 12}
b) R ∩ S = {4, 8, 12}
c) n(R) = 5
d) S' = {3, 7, 9, 13}
e) n(R') = 5
f) n(R ∪ S) = 8

Q3 a) [Venn diagram: ξ, C and D overlapping. C region: 8. Intersection: 4. D region: 9. Outside: 6]

b) i) $\frac{12}{27} = \frac{4}{9}$
   ii) $\frac{14}{27}$

Q4 a) A ∩ B = {4, 16}
b) [Venn diagram: ξ, A and B overlapping. A region: 6. Intersection: 2. B region: 2. Outside: 6]

c) i) $\frac{8}{16} = \frac{1}{2}$
   ii) $\frac{2}{16} = \frac{1}{8}$
   iii) $\frac{12}{16} = \frac{3}{4}$

Pages 70-71 — Graphs and Charts

Q1 a) Discrete.
b) [Bar chart: No. of Families (y-axis 0-3) vs No. of Children (x-axis 0-5). Values: 0→1, 1→2, 2→3, 3→1, 4→2, 5→1]

Q2 a) School A
b) School B
c) 10

Q3 a) £75
b)

Money For	Cost (£)	Degrees
Food	4800	64°
Accommodation Costs	3450	46°
Household Services	1800	24°
Leisure/Holidays	2400	32°
Insurance	900	12°
Clothes	1200	16°
Repairs and Decor	600	8°
Travel	2250	30°
Tax, N.I. and Pension	8100	108°
Savings	1500	20°

Answers

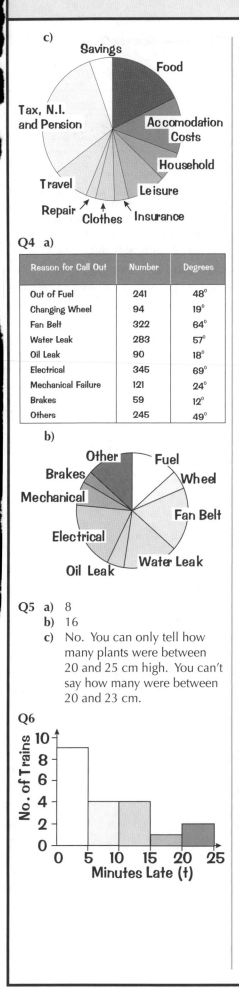

c)

Savings, Food, Tax, N.I. and Pension, Accomodation Costs, Household, Travel, Leisure, Repair, Clothes, Insurance

Q4 a)

Reason for Call Out	Number	Degrees
Out of Fuel	241	48°
Changing Wheel	94	19°
Fan Belt	322	64°
Water Leak	283	57°
Oil Leak	90	18°
Electrical	345	69°
Mechanical Failure	121	24°
Brakes	59	12°
Others	245	49°

b)

Other, Fuel, Brakes, Wheel, Mechanical, Fan Belt, Electrical, Oil Leak, Water Leak

Q5 a) 8
b) 16
c) No. You can only tell how many plants were between 20 and 25 cm high. You can't say how many were between 20 and 23 cm.

Q6

No. of Trains vs Minutes Late (t)

Q7

"Best Attraction"	Degrees on chart	No. of people
Neptune Rocket	70°	7000
Spooky Castle	50°	5000
Buccaneers' Boat	70°	7000
Zambezi Rapids	40°	4000
Death Canyon Railway	20°	2000
Supernova Experience	110°	11000

Page 72 — Mean, Median, Mode and Range

Q1 Mode = 3, Median = 3.

Q2 a) Mean = 6.9, Median = 8, Mode = 0, Range = 15.
b) Mean = 3.89, Median = 4, Mode = 3, Range = 4.
c) Mean = 8, Median = 8, Mode = 8, Range = 0.
d) Mean = 2.67, Median = 2, Mode = 2, Range = 3.

Q3 Mean = 239.3,
Median = 239,
Mode = 239.
The label on the box is pretty good – it says 240 maggots per box and the mean, median and mode are all within one of that.

Q4 Mean Pay = £4.52,
Median Pay = £4.63,
Mode Pay = £4.75.
Yes, looking at the mean, median and mode pay, Gary is right in thinking most of his friends earn more.

Q5 a) Mean = 4.08
b) 0 0 0 0 0 0 1 1 1 1 2 2 2 2 2 3 3 3 3 4 5 5 18 20 24.
Median = 2
c) 5 pupils above mean, 20 below it.

Pages 73-75 — Averages from Frequency Tables

Q1 Mean 14.3, median 14, mode 14, range 10

Q2 Mean 11.28, median 11, mode 10, range 8.

Q3 a)

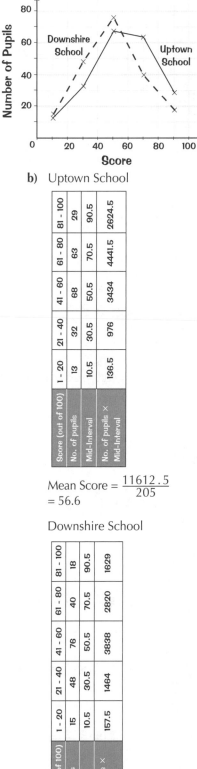

Number of Pupils vs Score. Downshire School, Uptown School.

b) Uptown School

Score (out of 100)	1 - 20	21 - 40	41 - 60	61 - 80	81 - 100
No. of pupils	13	32	68	63	29
Mid-Interval	10.5	30.5	50.5	70.5	90.5
No. of pupils × Mid-Interval	136.5	976	3434	4441.5	2624.5

Mean Score = $\dfrac{11612.5}{205}$ = 56.6

Downshire School

Score (out of 100)	1 - 20	21 - 40	41 - 60	61 - 80	81 - 100
No. of pupils	15	48	76	40	18
Mid-Interval	10.5	30.5	50.5	70.5	90.5
No. of pupils × Mid-Interval	157.5	1464	3838	2820	1629

Mean Score = $\dfrac{9908.5}{197}$ = 50.3

c) Not necessarily. There are other factors such as the pupils' ability, teaching, etc.

Answers

Q4 a)

Age (whole number of years)	Frequency	Mid-Interval	Frequency × Mid-Interval
0 - 9	107	4.5	107 × 4.5 = 481.5
10 - 19	130	14.5	130 × 14.5 = 1885
20 - 29	132	24.5	3234
30 - 39	144	34.5	4968
40 - 49	120	44.5	5340
50 - 59	113	54.5	6158.5
60 - 69	110	64.5	7095
70 - 79	90	74.5	6705
80 - 89	49	84.5	4140.5
90 - 99	5	94.5	472.5

$$\text{Mean} = \frac{40480}{1000} = 40.48$$
$$= 40.5 \text{ years old (1 dp)}$$

b) 30 – 39 yrs age group.
c) 30 – 39 yrs age group.

Q5 a)

Age (whole number of years)	Frequency	Mid-Interval	Frequency × Mid-Interval
0 - 9	272	4.5	1224
10 - 19	227	14.5	3291.5
20 - 29	173	24.5	4238.5
30 - 39	124	34.5	4278
40 - 49	99	44.5	4405.5
50 - 59	60	54.5	3270
60 - 69	29	64.5	1870.5
70 - 79	8	74.5	596
80 - 89	6	84.5	507
90 - 99	2	94.5	189

$$\text{Mean} = \frac{23870}{1000}$$
$$= 23.87 \text{ years old}$$

b) 20 – 29 yrs age group.
c) 0 – 9 yrs old.

Q6 a)

Wage (£)	No. of Employees	Mid-Interval	Frequency × Mid-Interval
0 ⩽ w < 5000	0	2,500	0
5000 ⩽ w < 10000	29	7,500	217,500
10000 ⩽ w < 15000	31	12,500	387,500
15000 ⩽ w < 20000	16	17,500	280,000
20000 ⩽ w < 25000	8	22,500	180,000
25000 ⩽ w < 30000	5	27,500	137,500
30000 ⩽ w < 35000	2	32,500	65,000
35000 ⩽ w < 40000	2	37,500	75,000

$$\text{Mean Wage} = \frac{1342500}{93}$$
$$= £14,400$$
$$\text{(nearest £100)}$$

b) £10,000 to £15,000 class.
c) £10,000 to £15,000 class.

Q7 a)

Wage (£)	No. of Employees	Mid-Interval	Frequency × Mid-Interval
0 ⩽ w < 5000	0	2,500	0
5000 ⩽ w < 10000	2	7,500	15,000
10000 ⩽ w < 15000	30	12,500	375,000
15000 ⩽ w < 20000	28	17,500	490,000
20000 ⩽ w < 25000	17	22,500	382,500
25000 ⩽ w < 30000	10	27,500	275,000
30000 ⩽ w < 35000	5	32,500	162,500
35000 ⩽ w < 40000	4	37,500	150,000
40000 ⩽ w < 45000	2	42,500	85,000
45000 ⩽ w < 50000	2	47,500	95,000

$$\text{Mean Wage} = \frac{2030000}{100}$$
$$= £20,300$$

b) £15,000 to £20,000 class.
c) £10,000 to £15,000 class.

Page 76 — Scatter Graphs

Q1 a)

b) Strong positive correlation
c) Approx £4600-£4800
d) Approx 360-380

Q2 a) and b)

c) Approx £17-£19
d) Approx 4-4.5 hrs

Q3 a)

b) The number of miles a car can travel per litre of fuel decreases as the engine size increases. Negative correlation.
c) Approx 9-10 miles per litre